T0201583

Philosophy of Nature

Philosophy of Nature
Paul Feyerabend

Edited with an Introduction by
Helmut Heit and Eric Oberheim

Translated by Dorothea Lotter
with assistance from Andrew Cross

polity

First published in German as *Naturphilosophie*, © Suhrkamp Verlag, Frankfurt am Main, 2009

This English edition © Polity Press, 2016

The translation of this work was supported by a grant from the Goethe-Institut which is funded by the German Ministry of Foreign Affairs.

Polity Press
65 Bridge Street
Cambridge CB2 1UR, UK

Polity Press
350 Main Street
Malden, MA 02148, USA

ISBN-13: 978-0-7456-5159-0

A catalogue record for this book is available from the British Library.

Library of Congress Cataloging-in-Publication Data

Names: Feyerabend, Paul, 1924-1994, author.
Title: Philosophy of nature / Paul Feyerabend.
Other titles: Naturphilosophie. English
Description: English edition. | Malden, MA : Polity, 2016. | Includes bibliographical references and index.
Identifiers: LCCN 2015042820| ISBN 9780745651590 (hardcover : alk. paper) | ISBN 0745651593 (hardcover : alk. paper)
Subjects: LCSH: Philosophy of nature.
Classification: LCC BD581 .F4413 2016 | DDC 113--dc23 LC record available at http://lccn.loc.gov/2015042820

Typeset in 10.5 on 12 pt Sabon by Servis Filmsetting Ltd, Stockport, Cheshire
Printed and bound in the United Kingdom by Clays Ltd, St Ives PLC

For further information on Polity, visit our website: politybooks.com

CONTENTS

PAUL FEYERABEND, AN HISTORICAL PHILOSOPHER OF NATURE

An Introduction by
Helmut Heit and Eric Oberheim

"An enthusiastic and very engaged student whose talents are far above average. At times he may give in to the urge to make impertinent comments." These remarks were entered on Feyerabend's report for the 1939/40 school year by teachers of Vienna's Public High School for Boys (*Staatliche Oberschule für Jungen*).[1] More than a few others would later record similar experiences in their interactions with him. Feyerabend was undoubtedly one of the most remarkable and controversial philosophers of science in the twentieth century, one who even drew attention outside the world of higher education. He had multiple interests and was interesting to many. While there is hardly any doubt about his above-average talents, assessments of his level of engagement are occasionally less enthusiastic. Feyerabend had a reputation for not being an excessively eager, committed, and thorough researcher, which was due at least in part to his nonchalant manner and snide remarks about the learned knowledgeableness of his colleagues. More than a few interpreted some of his remarks as unwelcome "impertinent comments," much as did his former teachers. And in the course of a general reckoning with certain relativistic and skeptical developments in philosophy of science, the journal *Nature* labeled Feyerabend the "Salvador Dali of academic philosophy and currently the worst enemy of science" (Theocharis and Psimopoulos

---

[1] This and other school transcripts with similar content are part of the Feyerabend collection at the Philosophical Archive of the University of Constance, archive no. PF 9-3-26.

1987: 596). However, according to the authors of that piece, in this respect Feyerabend was just a little ahead of Karl Popper, Thomas Kuhn, and Imre Lakatos, since philosophical reflection on science on the whole was undergoing an undesirable development. From that point of view Feyerabend had a place in the company of the classic figures of post-positivist philosophy of science. Of the "Big Four" in twentieth-century philosophy of science, however, it is Feyerabend to whom the label "enemy of science" has mainly stuck, first in a manner suitable for the media (Horgan 1993) and later, posthumously, as an ambivalent honorary title (Preston, Munévar, and Lamb 2000).

This volume of *Philosophy of Nature* is well suited to shed new light on Feyerabend's work and philosophical development as well as on his alleged hostility to science. In what follows we would like to introduce the reader to this text in three steps. (1) In the first step, after beginning with a brief initial summary of Feyerabend's philosophical development, we will reconstruct the history of this volume and give the reasons why its publication was delayed for more than thirty years. (2) In the second step, we will explore the special significance of the manuscript both for Feyerabend research and for our understanding of the development of our conceptions of nature. In *Philosophy of Nature*, Feyerabend presents himself as an interpreter of early Greek thought and a genealogist of Western rationalism. Thus, this text reveals a fascinating perspective not only on the history of philosophy of nature, but also on some hitherto little-explored aspects of Feyerabend's thought. At the same time, this work from the early 1970s constitutes a core resource for our understanding of the similarities and differences between the early and the late Feyerabend. It is the missing link for our understanding of Feyerabend's later radicalization and of its justification and its scope within the continuum of his thought. (3) The final part of the introduction provides an overview of the structure and contents of Feyerabend's *Philosophy of Nature*.

1. The History of an Unfinished Project

Feyerabend's philosophy has always been intimately connected with the scientific and philosophical discussions of his time, and he frequently participated directly in these discussions through his personal contacts. In the late 1940s, while studying in Vienna under Felix Ehrenhaft and Victor Kraft, he obtained some direct insights into the logical positivism of the Vienna Circle and its problems, which were of fundamental importance for the continuing development of philosophy of science on an international level. During this time, he also met Ludwig Wittgenstein, before deciding to accept an offer to work for Karl Popper in London. In the early 1950s, Feyerabend met with Niels Bohr a few times and became one of the most prominent philosophical critics of what would later be known as the Copenhagen interpretation of quantum mechanics. In 1962, along with Thomas Kuhn, his colleague at the time at the University of California, Feyerabend initiated the historical turn in philosophy of science, which subsequently adopted a stronger focus on the history and sociology of the sciences instead of regarding science exclusively as a logical system. During the 1970s, he became a strong critic first of Karl Popper's thought and his school of philosophy, and later of rationalism in a more fundamental sense. The catchphrase "Anything goes" from *Against Method* (Feyerabend 1975a, 1975b) drew attention both within and beyond academic philosophy. His subsequent works, especially *Science in a Free Society* (1978a, 1978b), "Science as Art" (1984a, 1984b), and *Farewell to Reason* (1987b, 1987c), were important contributions to a general discussion about the potentials and limits of the sciences in the Western world, as it was conducted in connection with postcolonial, postmodern, and ecological trends in the last third of the twentieth century. The potentials and limits of a scientific worldview are also themes in his final, posthumously published book, *Conquest of Abundance* (1999a, 1999b). Feyerabend's autobiography, *Killing Time* (1994a, 1994b), is a must-read, giving us insight into his dynamic life at the center of contemporary debates. And now *Philosophy of Nature* has revealed a hitherto little-known aspect of Feyerabend: the historical philosopher of nature and theoretician of the development of ancient philosophy.

While writing his major work, *Against Method*, whose first version was published in English in 1975, Feyerabend also worked on a comprehensive *Philosophy of Nature* in German. It was originally supposed to comprise three volumes reconstructing the history of human conceptions of nature from the earliest traces of Stone Age cave art to contemporary discussions of nuclear physics. Its working title was *Introduction to Philosophy of Nature* ("Einführung in die Naturphilosophie"). Since, however, it does not represent a genuine introductory work on the topic, but rather Feyerabend's own independent research, and thus is more of an historical reconstruction of the current situation than an introduction, we have decided to omit this misleading characterization from the published title of the work.

The project was not completed at the time; it was unknown in the late 1970s and apparently even forgotten by Feyerabend himself. The title continued to appear for a while in earlier bibliographies, but eventually disappeared.[2] In his autobiography, *Naturphilosophie* is not mentioned once. There are only a few isolated places in his later writings where it is mentioned at all. For example, Feyerabend mentions working on an introduction to philosophy of nature in a letter to Hans Albert (Baum 1997: 133). Yet the editor of the correspondence uses this reference primarily as evidence for Feyerabend's notoriously unreliable biographic and bibliographical statements: "Many projects were never realized; and even when a project was designated as 'in print' that does not mean it was actually published. For example, a book on philosophy of nature that had been scheduled to be published with *Wissenschaftliche Buchgesellschaft* was never released" (Baum 1997: 8). Baum was obviously unaware of the existence of the present work, which, however, at the time was scheduled to be released not by Wissenschaftliche Buchgesellschaft but by Vieweg in Braunschweig.

For this reason we were quite stunned when the uncompleted product of Feyerabend's efforts showed up during a research project at the Philosophical Archive of the University of Constance.[3] It soon became obvious that the 245-page photocopy of a typescript was an important new source for Feyerabend research. The text covers in five chapters the development of our human understanding of nature from the earliest cave paintings and records of early history through the Homeric *aggregate universe* to the *substance universe* of the Pre-Socratics, especially Parmenides. For the first time we had encountered a thorough discussion of the "rise of rationalism" in Greek antiquity, to which Feyerabend had repeatedly alluded. Later,

---

[2] A bibliography put together by Feyerabend himself and dated "April 1976" (PF 3-1-9) lists "Einführung in die Naturphilosophie,? Braunschweig 1974" as his sixth book (before *Against Method*). However, it is marked with red text marker and a handwritten note by Feyerabend, "never published." A somewhat later bibliography (PF 3-1-5) still mentions the "Naturphilosophie" as title no. 92: "Einführung in die Naturphilosophie und Mythenlehre," where the word "Darmstadt" has been struck through and replaced with a handwritten "Braunschweig 1976." But the "Naturphilosophie" was entirely absent shortly afterwards in a bibliography spanning the time until 1977 (PF 3-1-1).

[3] The material with archive no. PF 5-7-1 was discovered by Eric Oberheim and Torbjorn Gunderson in August 2004, when they were researching Feyerabend's literary estate following a seminar on *Against Method* at the Humboldt University of Berlin. They discovered the hitherto-unidentified book manuscript in a folder hidden under Feyerabend's dissertation "Zur Theorie der Basissätze" ("On the Theory of Basic Statements"; PF 5-6-2, Feyerabend 1951).

we happened quite by accident upon a reference to Feyerabend's unpublished *Naturphilosophie* in a work by Helmut Spinner.[4] It turned out that at the time Spinner was supposed to act as the editor of the three volumes and had already invested considerable time and effort in this project. We are grateful that he gave us access to these preliminary studies as well as to a second, more detailed version of the typescript comprising a total of 305 typewritten pages and including an additional sixth chapter. This sixth chapter contains an outline of the development of philosophy of nature from Aristotle to Bohr. The various chapters are executed with varying degrees of thoroughness, but overall they constitute a continuous and internally cross-linked argument. Unlike the fragmentary legacy of *Conquest of Abundance* (1999a, 1999b) they actually constitute a consistent, if not editorially completed, monograph.

Due to its not having been quite completed, however, the now-published *Philosophy of Nature* provides fascinating insights into this philosopher's workshop. It is especially suited to correct the image of a slightly airy thinker, which had been cultivated by Feyerabend himself. Though he employs an effortless writing style while still getting carried away with tart (and not always firmly justified) remarks, he does so in the context of a comprehensive discussion of the relevant contemporary material and an "enormous reading quota" (Hoyningen-Huene 1997: 8), which is clearly discernible in the work. In this book Feyerabend presents himself not only as an agent provocateur but also as an academic who has worked hard and studied a great deal of material. He aired his grievance about this in the following letter to Imre Lakatos of May 5, 1972:

> Dear Imre, Damn the *Naturphilosophie*. I do not have your patience for hard work, nor do I have two secretaries, a whole mafia of assistants who bring me books, check passages, Xerox papers and so on. If anarchism loses, then this is the most important reason. The examples which I find, are in books which *I* have found in the stacks myself, which *I* have carried myself, which *I* have opened myself, and which *I* have returned myself. [. . .] The very bloody version has been written by myself, never have I asked a secretary to do my dirty work.
>
> (Lakatos and Feyerabend 1999: 274f.)

We may speculate whether not only anarchism but also the "damned *Naturphilosophie*" might have failed due to an excessively

---

[4] Spinner also intended to use Feyerabend's thoughts toward a new interpretation of the development of western rationalism in antiquity and accordingly refers to Feyerabend's then unpublished work in two footnotes (Spinner 1977: 33 n. 99, 37 n. 121).

high workload. In any case, we know that Feyerabend dropped this project in the course of the late 1970s. His collaboration with Helmut Spinner ended in spring 1976 when Feyerabend apparently decided to undertake a substantial revision of the previous manuscript as well as his subsequent approach. This decision may have been in part due to Spinner's comprehensive comments and references, which Feyerabend appears to have valued and which encouraged him to revise the entire volume. At the same time, it appears that their collaboration was not entirely unproblematic, even though the agreements between Feyerabend, Spinner and publisher Vieweg were canceled by unanimous consent. This notion is supported by some public differences that occurred soon thereafter. For example, Spinner deplored Feyerabend's "philosophical idling" (Spinner 1977: 589), while the latter mocked Spinner's "illiteracy" (Feyerabend 1978b: 102). Nonetheless, in 1977 Feyerabend announced that he was planning to produce various publications "over the next two decades" in order to "remove some moral and intellectual garbage, so that new forms of life could appear. [. . .] This also includes my *Einführung in die Naturphilosophie*, which was supposed to be released in 1976 but which I have withdrawn in order to conduct some larger revisions" (Feyerabend 1977: 181). That he never implemented these plans is probably also due to the reactions to his other book from this period.

While before the mid-1970s Feyerabend was mainly a successful, argumentative, and respected philosopher of science, with his magnum opus *Against Method* he found himself catapulted into the center of contemporary intellectual and cultural debate. The predominantly negative reactions to *Against Method* may have had a two-sided effect on him with respect to his work on *Philosophy of Nature*. It is possible that it raised his standards of textual quality and clarity in order to prevent further misinterpretations. For though Feyerabend would rant and rave about "Sunday readers", "illiterates", and "propagandists" (1978a: 100ff.), he probably still felt responsible for those misinterpretations to at least some extent. This is also confirmed by the comprehensive edits that he repeatedly applied to the text. *Against Method* was originally published in 1970 as a long essay in *Minnesota Studies in the Philosophy of Science*. The first English-language book version was released in 1975. In the subsequent years Feyerabend used the two new editions (1988, 1993) as well as the two German translations (1976, 1983) to undertake comprehensive revisions and edits; hence we are today confronted with at least six versions of *Against Method*, which at times deviate significantly from one another with regard to content, scope, and argumentation.

Looking back, Feyerabend wrote in his autobiography, "*AM* is not a book, it is a collage" (1994a: 139). Though this collage established his international fame, it did not necessarily have a positive effect on his temper and self-esteem: "Somewhere in the middle of the commotion I grew rather depressed. The depression stayed with me for over a year [. . .] I often wished I had never written that fucking book" (1994a: 147). Feyerabend spent many years explaining *Against Method*. Perhaps his *Philosophy of Nature* would have been a better response to his critics than *Science in a Free Society*.[5] In any case, the debate surrounding *Against Method* and the related professional and personal strains were probably an important factor in Feyerabend's decision not to publish the present work despite the fact that it was near completion.

We have added a few more documents to the now-available book, which may give the reader further insights into the subsequent fate of *Philosophy of Nature*, as well as into Feyerabend's own assessment of his academic developments, achievements, and goals. A lengthy and informative letter written by Feyerabend back in December 1963 to Jack Smart constitutes a particularly interesting source regarding the history of the *Philosophy of Nature* project prior to the actual typescript. In it Feyerabend conveys to his Australian colleague that he had always wanted to write about the nature of myths in order to show that they are fully developed alternative worldviews. In doing so he combines various notions both from philosophy of language and from Kant, according to which conceptual schemes are always a factor in the constitution of our worldviews, with the notion that these schemata are neither innate nor historically invariant. Rather, historical research and the comparison of cultures both suggest the co-existence of alternative worldviews, which are equally fully developed, independent, and functional. Feyerabend then quotes from Nietzsche's "On Truth and Lies in a Nonmoral Sense" to illustrate that notion, using the liveliness and completeness of Greek myths as an example. He later reused the same quote and the same thought not only in his *Philosophy of Nature* (see chapter 3.3) but again 20 years later in his essay "Science as Art" (1984a, 1984b). In addition, the letter to Smart discloses a specific perspective on Feyerabend that can easily be overlooked in the later-published "Reply to Criticism: Comments on Smart, Sellars and Putnam"

---

[5] As Grazia Borrini-Feyerabend wrote in her Preface to *Conquest of Abundance*, in the last decade of his life Feyerabend was "not at all pleased with *Science in a Free Society*, which he did not want to see reprinted" (Feyerabend 1999a: xi).

(1965a) due to the large number of points discussed in that piece. Feyerabend's basic tendency to compare the scientific view of nature with mythical and ethnological alternatives dates back to the early 1960s, though initially his conclusions were less radical. This fact is confirmed also by two hitherto-unpublished texts that are very revealing both autobiographically and with regard to his philosophical development.

In his 1977 "Preparation" of a sabbatical year, Feyerabend talked about his increasingly skeptical approach to scientific rationality. Starting with historical investigations of actual scientific practices, he at first recognized the restricted validity of methodological rules. From there he eventually reached the point of fundamentally challenging the validity of any criteria for demarcating science from non-science. But it was only by exploring myths and early Greek art that he developed his thesis that there can be fully developed alternatives to a scientific worldview that at the same time cannot be evaluated on the basis of scientific criteria but only on their very own criteria. Eventually he realized that even the putative rules of reason are unable to make any essential distinction between science and non-science. For these reasons he was planning to work on a long-term project to develop a novel theory of knowledge that would account for this situation. A first step in this direction, his "short range plan," was revising and completing his *Philosophy of Nature*. Now, a later report on a sabbatical year, which was written in 1985, no longer mentions *Philosophy of Nature*, though it does still refer to the long-term and the short-term project. Aspects of the topics addressed reappear, especially those related to ancient mythology and worldviews, and the list of interlocutors included in the work report is also very informative. And yet it becomes obvious that Feyerabend completed neither the long-term nor the short-term project. However, in the editors' view this result should not be interpreted as an indication that Feyerabend – as may not be entirely uncommon in connection with requests for sabbaticals – did not seriously plan to fulfill the projects in the first place. Rather, the now-accessible *Philosophy of Nature* is proof of his sincerity with regard to this work and plan, for all of his notorious anarchistic self-staging. To the extent to which this text gives us evidence about Feyerabend's motives and questions in the 1970s, it also closes the gap between the putative earlier scientifically interested, serious philosopher of science and the later *enfant terrible* who is generally interested in cultural philosophy and critique of society.

2. *Philosophy of Nature* in the Context of Feyerabend's Philosophical Development

The special significance of *Philosophy of Nature* for the balance between continuity and change in Feyerabend's thinking can be understood only against the background of his earlier works. At first sight his works from the 1950s and 1960s appear to be fairly heterogeneous, as if there was no common organizing core. One could easily be tempted into reading them as a series of disconnected critical essays, developing partly contradictory ideas in various directions without being systematically linked in any way. This is hardly surprising; Feyerabend did, after all, consider himself an epistemological anarchist. Furthermore, he often had recourse to immanent criticism, seemingly adopting other authors' positions in order to bring out their internal problems. Consequently, his own standpoint, to the extent that he had one, often remained hidden. However, a more detailed look at Feyerabend's earlier works reveals the astonishingly exact repetition of a certain figure of thought consisting of two elements: the otherwise distinct objects of his criticism all appear as different forms of *conceptual conservatism*, and his criticism is always based on the presumption that there are hitherto unnoticed *incommensurable alternatives* to the prevailing notion. As early as in his dissertation "On the Theory of Basic Statements" Feyerabend used the idea – though not yet the full-fledged concept – of incommensurability to critically discuss conceptual conservatism in Heisenberg's concept of a closed theory. Feyerabend considered the conservative and exclusive use of established and successful concepts and theories to be problematic, since it illegitimately gives preference to existing theories over potential improvements, thereby obstructing scientific progress. This impulse can be found in almost all of his texts of the time. Feyerabend's early philosophy can be construed as a series of different attacks on any form of *conceptual conservatism*.[6] He pleaded for pluralism and theory proliferation to replace *conceptual conservatism*, which is very obvious in his previously mentioned "Reply to Criticism," e.g., criticism by thinkers such as Smart and Putnam:

> The main consequence is the *principle of proliferation: Invent, and elaborate theories which are inconsistent with the accepted point of view, even if the latter should happen to be highly confirmed and generally accepted.* [. . .] The theories which the principle advises us to

---

6 Oberheim (2005; 2006: esp. part II) offers more detailed reasoning in favor of this interpretation.

use in addition to the accepted point of view will be called the *alternatives* of this point of view.

(Feyerabend 1965a: 105f.)

Several aspects of this quote are notable and can give us more insight into Feyerabend's philosophical development as well as into the part played by his *Philosophy of Nature* in that process. First, it should be noted that, as he immediately makes clear by writing in a footnote, "when speaking of theories I shall include myths, political ideas, religious systems" (Feyerabend 1965a: 105). His concept of a theory, and hence of an alternative worthy of discussion, is not restricted to scientific systems of statements. Rather, it includes any construct of ideas that has an underlying range of applications, any comprehensive point of view that "is applicable to at least some aspects of everything there is." Thus, creation myths and speculative metaphysical systems are explicitly included. Second, Feyerabend in his 1977 "Preparation" of a sabbatical, printed for the first time in the present book, notably refers to his "Reply to Criticism" (Feyerabend 1965a) as summing up thoughts contained in "Explanation, Reduction, and Empiricism" (1962), "Problems of Empiricism" (1965b), and "Von der beschränkten Gültigkeit methodologischer Regeln" (1972a). These latter texts were his most important philosophical works at the time, introducing the concept of incommensurability to the contemporary debate in philosophy of science (in 1962, contemporaneously with Kuhn) and producing basic arguments for his critique of method. A central focusing point of *Philosophy of Nature* is the question of the extent to which a myth can qualify as a genuine, possibly incommensurable, alternative to a scientific theory. It is in the course of and through his work on this question that Feyerabend's assessment of the sciences becomes more radicalized. While in 1965 the goal was scientific progress, and theory proliferation the means, the later Feyerabend was less convinced that scientific progress is even desirable in every case. In a retrospective 1980 addendum in German to his earlier "Reply to Criticism" he wrote: "I am quite amazed when reading that radically scientistic treatise today. Though it opposes certain views about science such as extreme empiricism and monism, it still takes a pluralistically refined science as the basis for our approach to the world."

Feyerabend's change of mind, which he himself clearly recognized in retrospect, has been interpreted on occasion as a fundamental turn in his thought, and as a turn toward irrationalism.[7] He himself

---

[7] On his fundamental change of mind, see Preston (1997: 6ff., 139). Oberheim (2006: 281ff.) by contrast focuses on the continuity in Feyerabend's thought.

contributed not insignificantly to the appearance that this step toward a more radical relativization of the Western scientific worldview was primarily a consequence of his Berkeley experience. In his autobiography he refers to the student revolt and the opening of the universities to students with diverse cultural backgrounds, especially African Americans: "Should I continue feeding them the intellectual delicacies that were part of the white culture?" (Feyerabend 1994a: 123).[8] Thus Feyerabend's persistent doubts about the scientific worldview could be easily perceived as a socio-culturally motivated idiosyncrasy. However, if we add his *Philosophy of Nature* to the total picture of his development, we realize that not only does he present reasons for his skeptical stance, but these reasons should be regarded as a consistent expansion, a radicalization of his earlier philosophy. By implementing what he had announced in his letter to Smart, namely writing an essay on the phenomenon of myths as fully developed worldviews, as comprehensive theories in the sense described above, Feyerabend gets the chance to look at the scientific worldview as a whole and to confront it with an alternative: Homer's world. In accordance with his *principle of proliferation* only a comparison with such alternative views can create the possibility of serious testing and fair evaluation of the scientific worldview itself. We can describe the transition to Feyerabend's fundamental critique of science (which has occasionally been misunderstood as irrationalist) as a combination of his proliferation principle and his incommensurability thesis together with his research on antiquity. We can retrace this general radicalization beyond *Philosophy of Nature* in his interpretation of antiquity.

Throughout his life Feyerabend had a strong interest in topics concerning Greek antiquity. This can be traced back as far as to his early essay "Physik und Ontologie" (1954), which is also interesting from a philosophy-of-nature angle. Already in this essay he makes the distinction between "the mythological phase, the metaphysical phase, and the scientific phase" of explanatory worldviews (1954: 464). And it is still present in his *Conquest of Abundance*, on which he continued to work on his deathbed. He dedicated extended passages in his main work (1975a: 177ff.) to the transformation of thought in early Greek antiquity, which unfortunately went largely unnoticed, and discussed ancient sources in numerous of his other works. In his essay "Science as Art," he once described the underlying motive for

---

[8] Feyerabend's most explicit formulation of this motive for his increasing skepticism toward Western rationality can be found in the third revised English-language version of *Against Method* (Feyerabend 1975a: 163f. – see Hoyningen-Huene 1997: 8; Preston 1997: 4f.).

this as follows: "The introduction of abstract concepts in ancient Greece is one of the strangest chapters in the history of Western culture" (1984a: 50). For Feyerabend the transformation of thought in Greek antiquity may at first have been nothing but a particularly interesting case study, yet beyond this it also appears to be the basis for some central aspects of the Western concept of nature. Hence over and over again Feyerabend returned to his study of the early Greek intellectual world. However, his view of ancient thought underwent a fundamental change in the course of the 1960s, one that parallels his general development and is best understood as a radicalization. In his early texts Feyerabend's view of Pre-Socratic philosophy appears to have been strongly influenced by Popper. In *Knowledge without Foundations* (1961) he still assumed that scientific knowledge develops through a process of *conjectures and refutations*: "by a process of rational criticism which relentlessly investigates every aspect of the theory and changes it in case it is found to be unsatisfactory. The attitude towards a generally accepted point of view such as a cosmological theory or a social system will therefore be an attitude of criticism" (1961: 48). This approach basically replicates Popper's in "Back to the Pre-Socratics" (1958). But in the course of his partial disengagement from Popper, his assessment of the transformation of thought in Greek antiquity changed. Suddenly he saw the step from myth to logos no longer as an episode in a general history of progress driven by reasoned criticism of earlier positions. In his autobiography he cited his reading of Bruno Snell as a major factor for this change of mind.

> The long chapter on incommensurability [in *Against Method*] was the result of extended studies based mainly on three books: Bruno Snell's *Discovery of the Mind*, Heinrich Schäfer's *Principles of Egyptian Art*, and Vasco Ronchi's *Optics: The Science of Vision*. I still remember the excitement I felt when reading Snell on the Homeric notion of human beings.
>
> (Feyerabend 1994a: 140)

Until now this self-description by Feyerabend seemed less than plausible since, after all, he supported his arguments against conceptual conservatism and the forced application of a questionable scientific method mainly with examples from the early modern history of science. As a result, Feyerabend scholars have had little use for this passage.[9] In light of the present book, however, we may

[9] Helmut Heit (2007) has attempted a reinterpretation of the development of Greek philosophy partly on the basis of Feyerabend's considerations.

consider this retrospective assessment as a result of Feyerabend's work on his *Philosophy of Nature*, rudimentary and little-noticed traces of which can also be found in *Against Method* (and in later works). In *Philosophy of Nature* Feyerabend confronted the question of the origins of the Western scientific view of nature.[10] And here he first revealed his conviction that the Homeric-mythical worldview was defeated not by arguments but by history. Other references to the special significance of the ancient transformation of thought for Feyerabend's approach to science can be made accessible in a similar manner. In the revised German version of *Science in a Free Society* he explains that in *Against Method* he discussed three historical examples to illustrate the difficulties associated with methodologies in philosophy of science such as those proposed by Popper or Lakatos. Along with Einstein's replacement of classical mechanics and Galileo's defense of the Copernican system, "the third example [was] the transition from Homer's aggregate universe to the Pre-Socratics' substance universe" (1978b: 30). Though the last example belongs not to the history of science but rather to its prehistorical beginning, the "illustration of incommensurability that it provides is a close fit" (1978b: 30). The basic concepts of Homer's world and of the Pre-Socratics' world are incommensurable, for they "cannot be used simultaneously and neither logical nor perceptual connections can be established between them" (1975a: 169).[11]

It is important to keep in mind that for Feyerabend incommensurability was not the same as incomparability; rather, it merely indicates the lack of a common standard. Incommensurable theories cannot be internally related to one another; they can be compared only from a certain point of view, and this point of view should not be automatically viewed as superior, since its standards of comparison and evaluation are themselves always part of a worldview. Thus, to be incommensurable in this manner the Homeric and the Pre-Socratic worldviews need to be regarded as complete and fully functional conceptual and observational worlds. And this explains why Feyerabend in *Philosophy of Nature* presented an interpretation of Homeric epics and of archaic art and religion as constituting a universal theory with

[10] This is why at the time Helmut Spinner reasonably suggested that the text could also be titled *Introduction to Philosophy of Nature*, which would reflect the text as work on the origins of a specific philosophical reflection on nature in ancient Greece.
[11] It is, however, notable that in *Philosophy of Nature* the term "incommensurability" does not have this meaning at all; instead it is used only in its mathematical meaning. This may indicate that Feyerabend was more concerned with the development of genuine alternatives than with the concept of incommensurability.

empirical content. He confronted the naturalistic metaphysics and logocentric argumentation of the Pre-Socratics with the holistic and context-sensitive worldview of Homeric religion. This historical case also derives special significance from the fact that it does not actually belong to the history of science but rather marks an important aspect of its beginnings. In this way it makes it possible to confront science itself with an alternative. At the same time, evaluating non-scientific theories according to scientific standards is not legitimate. The special value of scientific standards would have to be demonstrated in a different way. It is only based on such considerations that the later Feyerabend could draw parallels between the belief in atoms and the belief in gods (1987b: 117f.) and demand a fair evaluation of the scientific worldview, something that had formerly been absent (1978b: 146; 1999a: 71f.).

All of this points to the fact that *Philosophy of Nature* was central in Feyerabend's philosophical development. In the 1950s and 1960s he challenged the various forms of conceptual conservatism by developing incommensurable alternatives to existing theories. The basis and objective of his criticism was the conviction, supported by arguments in philosophy of science, that only a pluralism of theories would not obstruct scientific progress. Investigating the ancient transformation from myth to logos, which is at the center of *Philosophy of Nature*, he focused on a specific case of incommensurable worldviews. This historical case is particularly notable since it marks the introduction of some general standards, positions, and values belonging to the Western scientific worldview. In particular, it is a preference for conceptual proof methods and abstract, context-independent thinking, as well as for a naturalistic metaphysics, that generally distinguishes scientific theories from their non-scientific alternatives. In the course of his work on *Philosophy of Nature* Feyerabend confronted the question of the uses and disadvantages of these standards for a happy life. The link between pluralism and progress continued to be of primary importance, but for the later Feyerabend scientific progress does not necessarily coincide with cultural and social progress. With his goal of contributing to a fair evaluation of the scientific worldview in mind, he presented myth and art as strong alternatives to this worldview. His later critical stance toward Western science thus turns out to be an extension of his criticism of conceptual conservatism through the development of incommensurable alternatives to existing theories; it follows the ideal of human progress.

3. Survey of the Course of Argument in *Philosophy of Nature*

In *Philosophy of Nature* Feyerabend has revealed himself to be a critical historian of Western theories of nature pleading for a pragmatic
use of human reason. He considered his work an "introduction"
inasmuch as it historically guides us to our present situation; it is a
genealogy of modern views of nature in light of past and possibly
even future alternatives. There were functionally successful alternatives to the modern scientific form of life, and like it they had both
advantages and disadvantages. Elaborating on the weaknesses of the
abstract scientific view of nature and the strengths of the alternative views, he considerably expanded the standard scope of such an
essay both historically and from an interdisciplinary point of view by
including three additional aspects. First, he examined prehistorical
and early historical periods as they are covered in research on Ice
Age art, Stone Age science, ancient Egyptian art and science, as well
as in Homer's world. Second, he discussed ethnographic and social
anthropological studies on indigenous tribes, challenging especially
the Eurocentric notion of primary or primitive thought, in order to
develop an adequate picture of mythical thought. Third, he included
classic art history in *Philosophy of Nature*. With that Feyerabend
exposed himself to the risk of occasional dilettantisms despite his
comprehensive reading list. Experts mostly reject some of his claims
today, such as the one about the fragmentary psychological state
of Homeric human beings.[12] But at the same time he extended the
traditional scope of historical studies in a most inspiring manner.
In this respect, his work is certainly superior to other introductions
to philosophy of nature, most of which fail to include Pre-Socratic
philosophy or even non-European cultures.[13]

In accordance with this comprehensive scheme Feyerabend dedicated the first two chapters to the earliest traces of natural science.
Based on archeological research and research in cultural history,
as well as on social-anthropological comparisons, he attempted a
reconstruction of the Stone Age conception of nature. In his theses

---

[12] The claim about the missing personal identity in Homeric human beings is mainly
due to Bruno Snell (1930; 1946a: 57–86, esp. 77f.; 1946b: 42–71). However, as
Bernhard Williams (1993: 88ff.) has shown, Snell's Kant-oriented concepts are insufficient to understand ancient concepts of subjectivity. Though the heroes in the Homeric
epics appear to have different body and world experiences from modern humans, they
still have individual names, consider reasons and motives, and act intentionally. See
Rappe (1995: 39f., 95f.) and Gill (1996: 29–40).
[13] See Gloy (1995), Mutschler (2002), and Esfeld (2004) as examples.

on Stonehenge as an early center of astronomy and on the Stone Age dynamic view of nature he relied heavily on a non-primitivist interpretation of the early cultures.[14] After all, Stone Age humans have the same biological and cognitive capacities as we do; they are fully developed members of the species *Homo sapiens*. Hence it is unlikely that their tools for understanding and controlling nature were structurally dysfunctional, resting on totally fantastic views. Feyerabend rejects the evaluation of historical material from the allegedly coherent and superior point of view of the present, finding it anachronistic and self-righteous. Instead, the traces of prehistoric and ancient cultures are just as theory-driven and partially successful conceptions of reality as our own; in the end, their quality can be evaluated only according to internal criteria. Accordingly, in the second chapter he interprets the Greek myths on the basis of a theory of nature myth: how does myth assist people in their understanding and control of nature? Also notable in this context is the detailed comparison of his own interpretation of myth with the works of Lévi-Strauss.

In the third chapter Feyerabend offers a realistic interpretation of both the archaic style in art and the Homeric epics. The manners of representation in archaic art are not due to a structural inability to create better paintings; rather, they adequately express a perceptual world that corresponds to them. Likewise, the structural characteristics of the Homeric epics are to be regarded as the consequence of a corresponding worldview. They exhibit a dynamic outlook paratactically composed of individual aggregates and responding flexibly in variable contexts. It is an open world that Feyerabend presents in a highly positive light. Not unlike those of Nietzsche and Whorff, his basic assumption is that our conceptual schemata play an essential part in the constitution of our perceptual world while at the same time being subject to historical changes. Scientific language and conceptions of reality are intertwined, and both are subject to possible changes.

Following these insights concerning the prehistorical period, *Philosophy of Nature* takes an interesting perspective on the transformation of our worldview in Greek antiquity, which at the same time aims at providing a better understanding of the Western conception of nature. The linguistic and empirical reality of the Homeric world dissolves and is replaced by the philosophers' world, as Feyerabend shows in chapter 4. This transition is shown to be a process not guided

---

[14] Some of the elements of his related theories were already familiar from his essay "Science as Art" (1984a: 25–9), which also contains various images that Feyerabend had collected for *Philosophy of Nature*.

by reason and rules, which means that it can be investigated only empirically by means of an emphatically historical analysis. Among the historical circumstances to be considered Feyerabend includes ceremonial and stylistic elements of the religious traditions, the effects of wars and periods of confusion, and also the hoplite phalanx, which undermines traditional heroic ideals. He places special emphasis on the impact of neighboring Eastern cultures. Furthermore, the critique of myths is already implicit in the myths themselves, as exemplified by Achilles' struggle for a new substantial concept of honor.[15] Feyerabend illustrates the novel features of the Pre-Socratic substance universe in the fifth chapter. The most notable characteristics of the thinkers of this time include giving priority to conceptual considerations over sensual experiences. According to Feyerabend, the early Greek philosophers' metaphysical presuppositions, especially the distinction between an allegedly simple and uniform reality and the merely apparent diversity in the world, denaturalize and dehumanize reality in favor of a dogmatic world of theory.[16] In this context he discusses Anaximander's cosmology, Xenophanes' criticism of religion and science, and especially Parmenides' substance universe. Most prominent are his frequently presented arguments for an alternative interpretation of Xenophanes' criticism of myths.[17] And yet he regards the Eleatic philosophy of Being in particular as the starting point of Western conceptions of nature due to its significant, and not always beneficial, influence: "On the contrary," he wrote in *Philosophy of Nature*, "a way of thinking such as Parmenides', that denies the existence of motion, thus leading Western thought astray for centuries, must strike us today as infantile and dreamlike."

In the sixth chapter Feyerabend outlines the way Western thought was led astray as well as how Western science, in his view, managed to retrace its steps to more holistic and dynamic concepts in the recent past. He often makes an effort here to develop a counter-position to the prevailing view, thus aiming to "strengthen the weaker argument"

---

[15] Feyerabend's brief reference in section 18 (ch. 4.3) to Achilles and Odysseus' exchange about honor was later developed into a detailed discussion in his *Conquest of Abundance* (1999a: 19ff.).

[16] Feyerabend expressed this criticism also in *Against Method* (1975a: 184f.), and it subsequently reappeared time and again in his later works. It is also the main topic of *Conquest of Abundance* (1999a); see Heit (2006).

[17] It is interesting that Feyerabend's criticism of Xenophanes became more radical in his later writings than in *Philosophy of Nature*, especially in that he later explicitly rejected Xenophanes' refutation of the Homeric gods on the basis of a *reductio ad absurdum* as a fallacy (Feyerabend 1984c; 1986; 1987a). In *Conquest of Abundance* his judgment of Xenophanes once again became a little more balanced, though without thereby relativizing the content of his criticism (Feyerabend 1999a: 49ff.).

(*ton hetto de logon kreitto poiein* – Protagoras, DK 80B6b). Thus he emphasizes the advantages of the Aristotelian conception of science, which in contrast to Descartes' mathematical approach to nature establishes a systematic connection between the theoretical and the practical elements of the scientific concept of nature. According to Feyerabend the most striking characteristic of early modern science is the lack of a genuine empirical foundation despite its accompanying empiricist rhetoric. In this context he defends Bacon from the accusation of an insincere empiricism, something that may be displayed in the naïve acquisitiveness of the Royal Society but is not reflected in Bacon and Galileo's ultimately theory-driven observations of nature. Perhaps most remarkable are Feyerabend's observations locating the roots of empiricism in Agrippa's occultist theories or in the experiments, which appear bizarre to us today, that were used to identify witches. In a detailed discussion of Hegel's observations Feyerabend suggests how a theory of concept dynamics is eventually reintroduced into Western thought. He closes by discussing the problems resulting from a concept of nature as a mere mechanism for the sciences and philosophy of science, using the examples of Newton, Leibniz, and Mach. He recognizes the beginning of a new, procedural, and once again more philosophical-mythological form of science, especially in David Bohm's physical theories. According to Feyerabend, the realization gradually sinks in that "There are more things in heaven and earth [. . .] Than are dreamt of in your philosophy" (*Hamlet*: I, 5). Understood correctly, this is not a plea for envisioning ghosts, but one for the flexible use of scientific philosophy with the awareness of its possibilities and limits. It is Feyerabend's objective in *Philosophy of Nature* to make explicit and promote the advantages of such an image of nature as well as of the corresponding more open-minded approach to science and its alternatives.

Note on Images in the Text

Feyerabend copied some of the images used in the text from other books, redrew others on the basis of originals, and designed the remainder himself. We replaced all copied images with printable scans, though not always from the original source that Feyerabend used. Simon Sharma artfully reconstructed Feyerabend's own drawings.

Acknowledgments

In concluding this introduction we would like to thank a number of institutions and individuals without whose support this work would likely have remained in the boxes belonging to the Feyerabend Collection in Constance. The German Research Foundation and the Humboldt University of Berlin made this edition possible with their financial and institutional support. The Carl and Max Schneider Foundation generously provided additional funding for a student assistant. Simon Sharma provided outstanding assistance with the transcription of the text and the editing of the numerous images. Without his numerous contributions this edition would not likely have been possible; thus we owe him special gratitude. We discussed parts of the text in various seminars with students, who assisted us in our verification of the sources as well as in our understanding of Feyerabend's thoughts. Prof. Helmut Spinner generously let us use his preliminary work and was amply available for information about the history of the project. We would also like to express special thanks to Sabine Hassel at the Humboldt University of Berlin as well as Brigitte Parakenings of the Philosophical Archive at the University of Constance. A heartfelt thank you goes to Grazia Borrini-Feyerabend, who joins us in our joy at being able finally to present this work of Feyerabend's to the public.

Berlin, October 2008
The Editors

Bibliography

Baum, Wilhelm (ed., 1997): *Paul Feyerabend–Hans Albert: Briefwechsel*, Frankfurt.

DK = *Die Fragmente der Vorsokratiker*, Hermann Diels and Walter Kranz (eds.), Berlin 1951.

Esfeld, Michael (2004): *Einführung in die Naturphilosophie*, Darmstadt.

Feyerabend, Paul (1951): *Per una teoria degli asserti di base* (Italian translation of Feyerabend's doctoral dissertation, "Zur Theorie der Basissätze" (On the Theory of Basic Statements)), trans. and ed. Stefano Gattei and Carlo Tonna, 2007.

– (1954): "Physik und Ontologic," *Wissenschaft und Weltbild: Zeitschrift für alle Gebiete der Forschung*, 7, 11/12 (Nov.–Dec.), pp. 264–76.

– (1961): *Knowledge without Foundations: Two Lectures Delivered on the Nelli Heldt Lecture Fund*, Oberlin.

– (1962): "Explanation, Reduction, and Empiricism," in: Paul Feyerabend, *Realism, Rationalism and Scientific Method: Philosophical Papers. Vol. 1*, Cambridge, 1981, pp. 44–96.

– (1965a): "Reply to Criticism: Comments on Smart, Sellars and Putnam," in: R. Cohen and M. Wartofsky (eds.), *Proceedings of the Boston Colloquium for the Philosophy of Science 1962–1964: In Honor of Philip Frank*, Boston Studies in the Philosophy of Science, vol. 2, New York, pp. 223–61.

Reprinted in: Paul Feyerabend, *Realism, Rationalism and Scientific Method: Philosophical Papers*. *Vol. 1*, Cambridge, 1985, pp. 104–31.
- (1965b): "Problems of Empiricism," in: Robert G. Colodny (ed.), *Beyond the Edge of Certainty: Essays in Contemporary Science and Philosophy*. *Vol. II*, Pittsburgh, pp. 145–260.
- (1972a): "Von der beschränkten Gültigkeit methodologischer Regeln," in: Paul Feyerabend, *Der wissenschaftstheoretische Realismus und die Autorität der Wissenschaften*, Braunschweig, 1978, pp. 205–48.
- (1972b): "On the Limited Validity of Methodological Rules," in: Paul Feyerabend, *Knowledge, Science and Relativism*, ed. John Preston, Cambridge, 2008, pp. 138–80.
- (1975a): *Against Method: Outline of an Anarchistic Theory of Knowledge*, London/New York, 1993.
- (1975b): *Wider den Methodenzwang*, trans. Herrmann Vetter, Frankfurt, 2nd rev. edn., 1983.
- (1977): "Unterwegs zu einer dadaistischen Erkenntnistheorie," in: Hans-Peter Duerr (ed.), *Unter dem Pflaster liegt der Strand (4)*, Berlin, pp. 9–88.
- (1978a): *Erkenntnis für freie Menschen*, Frankfurt, rev. edn., 1980.
- (1978b): *Science in a Free Society*, New York.
- (1984a): *Wissenschaft als Kunst*, Frankfurt.
- (1984b): "Science as Art: A Discussion of Riegl's Theory of Art and an Attempt to Apply it to the Sciences," *Art and Text*, 12/13 (Summer 1983/Autumn 1984), pp. 16–46.
- (1984c): "Xenophanes: A Forerunner of Critical Rationalism?" trans. John Krois, in: Gunnar Anderson (ed.), *Rationality in Science and Politics*, Boston Studies in the Philosophy of Science, vol. 79, Dordrecht/Boston/Lancaster, pp. 95–109.
- (1986): "Eingebildete Vernunft: Die Kritik des Xenophanes an den Homerischen Göttern," in: Kurt Lenk (ed.), *Zur Kritik der wissenschaftlichen Rationalität: Zum 65. Geburtstag von Kurt Hübner*, Freiburg/Munich, pp. 205–23.
- (1987a): "Reason, Xenophanes and the Homeric Gods," in: Paul Feyerabend, *Farewell to Reason*, London/New York, pp. 90–102.
- (1987b): *Farewell to Reason*, London/New York, 1988.
- (1987c): *Irrwege der Vernunft*, trans. J. Blasius, Frankfurt, 1990.
- (1994a): *Killing Time: The Autobiography of Paul Feyerabend*, Chicago.
- (1994b): *Zeitverschwendung*, trans. Joachim Jung, Frankfurt.
- (1999a): *Conquest of Abundance*, Chicago.
- (1999b): *Die Vernichtung der Vielfalt: Ein Bericht*, trans. Volker Böhnigk and Rainer Noske, Vienna, 2004.
Gill, Christopher (1996): *Personality in Greek Epic, Tragedy, and Philosophy: The Self in Dialogue*, Oxford.
Gloy, Karen (1995): *Das Verständnis der Natur. Erster Band: Die Geschichte des wissenschaftlichen Denkens*, Munich.
Heit, Helmut (2006): "Paul K. Feyerabend: Die Vernichtung der Vielfalt. Ein Bericht. Wien 2005," *Zeitschrift für philosophische Forschung*, 60/4, pp. 615–19.
- (2007): *Der Ursprungsmythos der Vernunft: Zur philosophischen Genealogie des griechischen Wunders*, Würzburg.
Horgan, John (1993): "Profile: Paul Karl Feyerabend: The Worst Enemy of Science," *Scientific American*, (May), pp. 36–7.

Hoyningen-Huene, Paul (1997): "Paul K. Feyerabend, *Journal for General Philosophy of Science*, 28, pp. 1–18.

Lakatos, Imre and Paul Feyerabend (1999): *For and Against Method: Including Lakatos' Lectures on Scientific Method and the Lakatos–Feyerabend Correspondence*, Chicago.

Mutschler, Hans-Dieter (2002): *Naturphilosophie*, Stuttgart.

Oberheim, Eric (2005): "On the Historical Origins of the Contemporary Notion of Incommensurability: Paul Feyerabend's Assault on Conceptual Conservatism," *Studies in History and Philosophy of Science*, 36, pp. 363–90.

– (2006): *Feyerabend's Philosophy*, Berlin/New York.

Popper, Karl R. (1958): "Back to the Presocratics," in: Karl Popper, *Conjectures and Refutations: The Growth of Scientific Knowledge*, London/New York, 2014, pp. 183–206.

Preston, John (1997): *Feyerabend: Philosophy, Science and Society*, Cambridge.

Preston, John, Gonzales Munévar, and David Lamb (eds., 2000), *The Worst Enemy of Science? Essays in Memory of Paul Feyerabend*, Oxford.

Rappe, Guido (1995): *Archaische Leiberfahrung: Der Leib in der frühgriechischen Philosophie und in außereuropäischen Kulturen*, Berlin.

Shakespeare, William [*Hamlet*]: *Hamlet*, Arden Shakespeare, Third Series, London, 2006.

Snell, Bruno (1930): "Das Bewußtsein von eigenen Entscheidungen im frühen Griechentum," in: Bruno Snell, *Gesammelte Schriften: Mit einem Vorwort von Hartmut Erbse*, Göttingen, pp. 18–31.

– (1946a): *Die Entdeckung des Geistes: Studien zur Entstehung des europäischen Denkens bei den Griechen*, Hamburg (2nd amended edn.).

– (1946b): *The Discovery of the Mind: The Greek Origins of European Thought*, trans. T. G. Rosenmeyer, New York, 1960.

Spinner, Helmut F. (1977): *Begründung, Kritik und Rationalität. I: Die Entstehung des Erkenntnisproblems im griechischen Denken und seine klassische Rechtfertigungslösung aus dem Geiste des Rechts*, Braunschweig.

Theocharis, T. and M. Psimopoulos (1987): "Where Science Has Gone Wrong," *Nature*, 329 (October 15), pp. 595–8.

Williams, Bernhard (1993): *Shame and Necessity*, Berkeley/Los Angeles.

EDITORIAL NOTES

The present edition of Paul Feyerabend's *Philosophy of Nature* was compiled based on two partly deviating typescripts. Both were photocopies of loose pages covered with typewritten text as well as handwritten corrections and additions. The original has been lost. One of the versions, the Archive Version, is from the Paul Feyerabend collection at the Philosophical Archive of the University of Constance (archive no. PF 5-7-1) and consists of 245 loose pages, which Feyerabend had divided into five chapters. According to a note on page 1a, it was written in 1971 and revised in 1976. The second version, the Spinner Version, is in the possession of Prof. Dr. Helmut Spinner, who at the time was supposed to act as the editor of *Philosophy of Nature*, which had been planned as a three-volume publication. A copy of this typescript can now be viewed at the Constance Archive as well. The Spinner Version is dated August 1974 and comprises 305 loose pages. The additions mainly relate to the additional sixth and the brief seventh chapter, but to some extent also to revisions in various other places in the text. Nonetheless the Spinner Version is not always the more up to date, since Feyerabend's 1976 revisions were made not to this version but to the older Archive Version. Thus, the textual basis of the present edition consists of two partly different versions that also had different revisions made by Feyerabend and display traces of Spinner's editorial work as well. We have generally used Feyerabend's most recent versions for the present edition, but we have not included the major additions by Spinner, which were in part content-related.

This book aims to make a reliable readers' edition available to Feyerabend scholars as well as to a larger audience interested in Feyerabend's reconstruction of philosophy of nature. Thus we did not include a comprehensive critical historical apparatus or a transparent reconstruction of the various stages of revision. We silently corrected obvious errors, while additional revisions of the text are documented in editors' footnotes. In particular, we had to perform some major rearrangements in order to turn the incomplete typescript into a readable book without altering the content or leaving any parts out. Feyerabend's original text did not contain any footnotes; instead it began with a note:

> I would like to suggest to my readers that they first read the essay straight through, leaving the passages in small print for later. In this way readers will obtain an overview of the underlying ideology. The text in small print contains further material, additional arguments, and bibliographical references. Occasionally it takes up an ancillary line within an argument and develops it further.
>
> (PF 5-7-1, p. 1)

We were unable to distinguish passages in small print clearly, especially further on in the text, not least due to Feyerabend's use of different typewriters (a consequence of his flexible lifestyle) as well as to various traces of editing that had accumulated over the course of several years. Furthermore, the repetitive bibliographical references at the end of argumentative passages were considerably detrimental to the flow of reading. Thus, the attempt to identify and leave passages in small print where they are in the text would have been contrary to Feyerabend's objective of providing readers with "an overview of the underlying ideology." This is why we moved the references and ancillary lines of arguments into an apparatus of footnotes, just as Feyerabend himself used to do in those publications that he supervised in person. In doing so we assigned the footnotes to the thematically corresponding passages.

Feyerabend structured the text into seven chapters and 43 consecutive sections. He used the section numbers, which are placed in square brackets in this edition, for cross-references within the text. Since, however, the sections vary greatly in length, do not always cover just one single topic, and are not always clearly marked off from subsequent sections (as, for example, with sections [5] and [6]), they are of limited usefulness for orientation within the text. For this reason we decided also to divide up the text according to subchapters, which we did on the basis of content-related criteria, and to introduce each subchapter with a suitable subchapter heading. Another considerably more significant structural change in the typescript was also performed with the aim of improving its readability. In *Philosophy of Nature* Feyerabend quoted from roughly 300 different works and included the references, which differ in their degree of thoroughness, immediately after the quotes inside the text. We replaced these references in the continuous text with author/year or, for classic texts, author/abbreviation of title, and we added a bibliography that lists all the literature used by Feyerabend. Furthermore, we added additional captions to the images in the text. It was not always possible to obtain reproducible originals from the same sources that Feyerabend used. In these cases, we had recourse to other resources and referenced the new sources. Our associate Simon Sharma digitally reconstructed drawings by Feyerabend.

To the extent to which it was possible, we reviewed all sources used by Feyerabend and supplemented the references as needed. Belying his later reputation, Feyerabend's quotes and references were accurate in most cases. Minor typos, such as the misrepresentation of the name of Hermann Neuwaldt (1984) as "Hermann Neustadt" or the wrong page number for a passage from Herodotus (*Histories* IV. 252 instead of IV. 152), were comparatively rare and have been silently corrected. We proceeded in the same way with any errors that may have occurred when Feyerabend copied text passages from a source, as in Olof Gigon's "Die Wandelbarkeit des Standpunktes, die der Elegie seit Archilochos eigentümlich ist, muss notwendig auch der Philosophie des Xenophanes einen besonderen Charakter gegeben haben" (1968: 157), where Feyerabend got confused about the tense while copying the text ("eigentümlich war" instead of "eigentümlich ist") and also forgot to copy the word "auch" ["also"]. *When checking the sources we strove as far as possible to use the same editions as Feyerabend.** Wherever this was not possible we have listed in the bibliography the sources that we used to check the reference in question. In some cases we were unable to check a reference and relied on the accuracy of Feyerabend's information. The bibliographical references in the sixth

---

* [Translator's note: Translations of German quotes in the Feyerabend text mostly follow already published English standard translations of the originals, where they were available. Where these were not available I translated directly from the (editor-corrected) quotes in the text, provided that these were originally in German rather than Feyerabend's translations of contents originally in other languages. I did find a number of inaccuracies in some of Feyerabend's German translations of English sources, which were corrected by simply copying the English original. In one case Feyerabend had apparently memorized a sentence from an English source that did not actually exist in the source, although the meaning of the sentence was broadly reflected in the contents of the text. In such a case I translated Feyerabend's own version of the quote and added a translator's note to clarify the inaccuracy.]

** [Translator's note: This English edition uses English standard translations of the classics rather than the German translations with which Feyerabend worked.]

chapter especially have a somewhat lower degree of completeness, and in some cases Feyerabend neglected to provide a reference at all. Wherever it was possible we added the missing reference without specifically noting this.

Since Feyerabend had teaching positions in the USA (Berkeley and Yale), Europe (London and Berlin), and in 1972 and 1974 even in New Zealand, he moved primarily in English-language environments during his work on *Philosophy of Nature*. This is noticeable from a few textual notes in English as well as some anglicisms in his use of German. For example, in *Philosophy of Nature* he often formed an adjective's comparative in the English style, by speaking of "mehr abstrakt" rather than "abstrakter" ["more abstract"], which would have been the correct form in German. In those cases in which Feyerabend did not use an existing German translation of sources in foreign languages, he referenced the original foreign-language source while providing his own German translation in the text. Due to external circumstances he also used different editions of the same source. This is most obvious with regard to the writings of Lévi-Strauss, whose *The Savage Mind* was quoted by Feyerabend in its original French as well as in German and English translation. Feyerabend's own German translations, of course, deviated from the published German standard translations, as the former were based on an English translation of the text. In some, though not all, cases Feyerabend later replaced his own translation with Hans Naumann's German translation of *The Savage Mind*, published by Suhrkamp. Just as with other revisions, we used Feyerabend's last version as the basis for *Philosophy of Nature*. His own translations reflect his interpretation and are usually accurate enough that we did not undertake any amendments, even though alternative interpretations would have been possible. We chose to proceed this way even in cases where Feyerabend had read a text that was originally in German in an English translation and then translated it back into German, as is the case with a passage from Martin Luther's *On the Babylonian Captivity of the Church* (1520). Luther's original text reads: "Denn was ohne Schriftgrundlage oder ohne erwiesene Offenbarung gesagt wird, mag wohl als eine Meinung hingehen, muss aber nicht notwendig geglaubt werden," and the English standard translation reads: "For that which is asserted without the authority of the Scripture, or of proven revelation may be held as an opinion, but there is no obligation to believe it." And Feyerabend turned this into: "Denn was ohne Autorität der Schrift und bewiesener Offenbarung behauptet wird, kann als Meinung gelten, man ist aber nicht verpflichtet es zu glauben." Even in such cases, in *Philosophy of Nature* we preserved Feyerabend's own text and included a reference to the original only when there were major errors.

Feyerabend entered Greek letters by hand into the typescript and they were not always easy to read, which is why we decided generally to render them in Latin letters to improve their legibility. Regarding the ancient authors we were not always able to identify the translations used by Feyerabend. In such cases we reference the standard translations in the bibliography, using the standard abbreviations of titles. It is possible, though not certain, that Feyerabend produced his own translations from Ancient Greek into German. Alternatively, he may have translated an existing English translation into German. In Vienna, Feyerabend had attended the science- and mathematics-oriented program at the Public High School for Boys and graduated with the highest grade in English and Latin, but he did not receive any Greek lessons at the school. However, he apparently began his own studies in Ancient Greek starting from the mid-1960s. According to Bert Terpstra, the editor of Feyerabend's posthumously published book *Conquest of Abundance*, Feyerabend produced many of the translations from Greek in that book himself (Feyerabend 1999a 45). In any case, Feyerabend did have a basic knowledge of the Greek language and was able to entertain thoughts about possible alternative translations. Wherever he did not provide translations of Greek words in the text we added such translations in the footnotes.

Philosophy of Nature
Paul Feyerabend

Preliminary Note

I would like to suggest to my readers that they first read the essay straight through, leaving the passages in small print for later. In this way readers will obtain an overview of the underlying ideology. The text in small print contains further material, additional arguments, and bibliographical references. Occasionally it takes up an ancillary line within an argument and develops it further.[1]

This essay is an "introduction" in the sense of guiding us historically toward our contemporary situation. The three forms of life that will be discussed – myth, philosophy, and science – cannot be strictly distinguished from one another, nor will we always encounter them in that order. Myth anticipates science while science has certain traits of myth. Philosophy, science, and myth sometimes live peacefully side by side, but at other times they deny one another the right to subsist. "Superstition" and "preconceptions" can be encountered everywhere. Moreover, science does not always have an advantage over myth and philosophy. It is interesting that in science, myth, and philosophy we have three different ways of comprehending the world,

---

[1] [Editors' note: In the course of this essay, Feyerabend does not keep up with the division of the text into a main argument and ancillary lines that he outlines here. In addition, the print image of the original typescript rarely made it obvious what was to be regarded as a passage in small print. In order to enable the reader to grasp the "underlying ideology" in one reading we chose a system of footnotes rather than different font sizes to accommodate the ancillary lines of argument and bibliographical references. Furthermore, the typescript was structured rather unsystematically into sections [1]–[43], on which Feyerabend occasionally relied for cross-references. We have supplemented this structure by introducing section headings. For further details see the Editorial Notes.]

which are *complete* in that they include both *objectives* and *methods* to achieve these objectives as well as *criteria* to determine success and failure. Ideally, the *transitions* from one state to the next, which for now we can reconstruct in detail only for the occidental part of the world, bring with them not just new *ideas* and *methods* but also new *objectives* and new *perceptions*, which unite to form new perceptual worlds.

Only one of these transitions will be discussed in more detail in the present essay. It is the transition from Homer's "aggregate universe" to the "substance universe" of Greek philosophy and science. There are historical reasons why this transition is relatively clear and distinct, so that it can be reconstructed in detail. There are many different factors having an impact on these early thinkers, which makes it necessary to move beyond the area of philosophy and include in the discussion the art, literature, and religious thought of the time as well. Such more comprehensive approaches reveal that "progress" in one area is almost always accompanied by regress in others. The transition to the philosophers' "rational" universe, in particular, bore problems that are still waiting to be solved today and that may be altogether unsolvable; a return to some form of myth may be advisable.

My objective is achieved when readers realize that a decision for or against "myth," or for or against a certain method in philosophy or science, is not easy and that there are always arguments in favor of both sides; and when they become familiar with some of the factors that commonly play into such decisions. The epistemological position at the basis of the present work is further explained in my essay *Against Method* (1975). The present text was originally composed in 1971 and revised in 1976 in light of new literature.

— 1 —

PRESUPPOSITIONS OF MYTHS, AND THE KNOWLEDGE OF THEIR INVENTORS

[1] We can reconstruct the development of views of nature in three different ways. *Archeology* familiarizes us with material products that enable us to draw conclusions about the ideas and knowledge available to the associated cultures. *Mythology* in the broad sense – that is, the study of legends, fairytales, rituals, sayings, esotericisms, chants, epics, and dreams – enables us to identify and partially decode fragments of bodies of archaic knowledge supplementing those indirectly discovered ideas. Finally, *comparative cultural anthropology* teaches us how contemporaneous preliterate small societies link ideas, "facts" of nature, social conditions, artifacts, and so on, and offers an analogy to supplement the findings of archeology and mythology.

To collect and interpret their rich and enigmatic material, these three basic disciplines rely on assistance from all of the other sciences, that is, from astronomy, biology, chemistry, physics, geography, etc. We need these branches of knowledge to date and analyze material (such as the origin and conditioning of the materials used in the creation of pieces of art, buildings, jewelry, and so on), and even more to *interpret* the information thus obtained. For how could we understand the astronomical information conveyed in a myth, or the astronomical function of a building, if we lacked precise knowledge of the events happening in the starry sky? Nor does a mere rudimentary mastery of auxiliary sciences suffice. The assumption that humans of the Stone or Bronze Age could have had only the most primitive knowledge of nature may be flattering to our progressivist self-image. But it has little plausibility, since Stone Age humans were already fully developed members of the species *Homo sapiens*, and it is incompatible with recent research. The environmental and societal problems that the early *Homo sapiens* had to face were incomparably

5

greater than the challenges facing our contemporary scientists. These problems had to be solved with the most primitive means, often without any division of labor or specialized skills, and the solutions arrived at indicate a level of intelligence and sensitivity that is clearly not inferior to ours.

For the longest time, anthropology and related disciplines maintained a terminological distinction between "barbarians," "savages," or "primitive peoples" on the one hand, and "civilized," "sophisticated," or "advanced" peoples, on the other. This terminology originates with the fairly primitive evolutionary notions of the nineteenth century, according to which a linear evolutionary development in the animal kingdom (which today is very much in doubt) is reflected in an equally linear evolution of human skills and cultures. Evans-Pritchard has revealed the pseudo-empirical nature of evolutionist theory in very clear terms. The evolutionists proceed by stipulating specific lines of evolution, which are then illustrated in a large volume of material collected in a way that is anything but impartial: "In spite of all their talk about empiricism in the study of social institutions the nineteenth-century anthropologists were hardly less dialectical [...] than the moral philosophers of the preceding century" (Evans-Pritchard 1964: 41). The empirical foundation drifts away even further when one considers that the missionaries who supplied Frazer and Tylor with reports often were enthusiastic supporters of the evolutionist theory. The theory entirely determined the questions they asked, the questionnaires they put together, and the choice of material to be collected.[2] In addition, they stipulated an evolutionary stage at which humans were still entirely natural beings untouched by the blessings of culture. Research has not confirmed that such an evolutionary stage ever existed, and even *a priori* its existence at any time is quite unlikely. Rousseau, who is often quoted as a witness to evolutionist theory, explicitly rejected the notion of such a stage. He "directed his energy toward discovering not a state of nature without culture but a culture that would realize man's true nature" (Gay 1970: 95). Rousseau's prize essay of 1750 on the question "Has the restoration of the sciences and arts contributed to the purification of morals?" as well as his "Last Reply" to M. Charles Bordes confirm this: "[A]ll barbarous Peoples, even those that are without virtue, nevertheless always honor virtue, whereas learned and Philosophic Peoples by

---

[2] On the problem of the "Concept of the Archaic in Anthropology," see Lévi-Strauss (1958: ch. VI). On the impact of evolutionist theory on anthropology, see Lowie (1937: chs. III and VI) as well as Evans-Pritchard (1964: chs. II–V).

dint of progress eventually succeed in turning virtue into an object of derision and despising it" (Rousseau, *Reply*: 67). Social unions, social collaboration, and discoveries and their dissemination occur even in the animal kingdom – for example, in the chimpanzee communities that Jane Goodall observed (Goodall 1971).[3]

The difference between nature and culture (wild versus tame; raw versus cooked; adorned versus unadorned), which even modern philosophers tend to pass lightly over,[4] was for many "savages" a *problem* they tried to understand through myths and to solve through rituals.[5] Occasionally a tribe would fall into the opposite extreme and reject "natural" functions with every indication of disgust.[6]

The opposition between *primitive* peoples and *civilized* peoples is to a large extent due to an overestimation of *written* language. Written language is often, but not always, linked to progressiveness (or greater intelligence). The languages of preliterate tribes often exceed the *civilized* languages familiar to us with respect to complexity. For this reason they cannot always be learned by adult Westerners (although children might still possess a certain talent to master them). We know of preliterate cultures with accomplishments comparable to the accomplishments of contemporary literate ones. We know that unrivaled pieces of art such as the *Iliad* and the *Odyssey* were composed and passed down by preliterate performers[7] and that the Rabbinical traditions, the commentaries on the Torah prior to the fifth century, were not written down but were passed from one generation to the next by professional memory artists. The Greeks, surely not the least intelligent people on Earth, were opposed to the introduction of written language into literature, and even Plato opposed purely written expositions of philosophical problems.[8]

A situation like this urgently demands the introduction of a new terminology. In this essay I speak of peoples of culture and peoples of reason, industrial peoples and pre-industrial peoples, literate peoples and preliterate peoples. That is, I do not assume an ideal state (which,

---

3 See also Lorenz's groundbreaking research, as for example in "The Companion in the Bird's World" (1935b), as well as Ardrey's more general and recent account (1967).
4 See the critical exposition in ch. V of Popper (1945: 57–85).
5 See Lévi-Strauss (1955: 203, 204, 214, 216; 219, 224) as well as the myths in Lévi-Strauss (1964) that deal with the evolution of cultural institutions in the natural world. Griaule (1965) – close connectedness of culture and *language* as well as a detailed theory of the development of the latter – as well as Kirk (1970: ch. IV, esp. 144f.), are more cohesive.
6 Cf. the Caduveo attitude toward the act of procreation and its effects, in Lévi-Strauss (1955: 162, 170).
7 See the report in Dodds (1951: 13ff.) as well as Page (1959) and Kirk (1965).
8 Plato (*Phaedrus*: 275a); see the discussion in Friedländer (1954: 114–32).

"of course," would be the one realized in the West) to use as a basis for my distinctions; rather, my distinctions are based on relevant structural differences between the societies described. In each case the details become clear from their context. Yet even an "objective" baseline study using "neutral" questions remains incomplete and misleading. It does not give respondents the opportunity to use arguments to explain and defend seemingly nonsensical parts of their myths and so leads to an incorrect assessment of the myths and their followers. This is not at all the way to find out about the argumentative skill of indigenous peoples. It took researchers such as Evans-Pritchard, who treated the members of the examined tribes as normal people and thus attempted to use arguments to convert the "savages" to their own Western scientific ideology, to realize that such "conversion" is not an easy matter. Though the members of the tribe at first had no idea of the structure and function of critical arguments, they quickly caught on – if not very enthusiastically – and soon turned rationalism against their very teachers. The arguments were met with smart counterarguments, and it turned out that even the seemingly most absurd form of life possesses or can be given a strong rationalist core, if its members so desire: "I found this as satisfactory a way of running my home and affairs as any I know of" (Evans-Pritchard 1937: 270), wrote Evans-Pritchard about the Azande, who did detailed oracle consultations prior to performing even the most trivial acts. Last, but not least, many of the tribes examined were in a state of dissolution at the time of the study: the "primitive" traces found in them were a result of the catastrophic influence of Western intruders (Lévi-Strauss 1958: ch. VI). What would New York look like if gas, electricity, and fuel were suddenly suspended?[9]

---

[9] On the role of evolutionist thoughts in depth psychology see Jung's "Concerning the Two Kinds of Thinking" (1916: 9–36). Jung introduced the *analysis of dreams* in his *Psychology and Alchemy* (1944b) as a method of mythology. His argument is that the soul contains impressions of eras long gone, that Christianity has not, or only very little, changed these impressions, and that it is therefore possible to reveal archaic material through psycho-archeology. See also Kerényi's prolegomena to Jung and Kerényi (1951a, 1951b) and the critical remarks in section [3] [*chapter 1.1*] of the present book. The following works may serve as an introduction to the problems of archeology: Burkitt (1963), Childe (1956), and Daniel (1967); they are historical and contain valuable excerpts from the works of earlier writers. I can also recommend Ceram (1949a, 1949b, 1957). Marshack (1971) offers a comprehensive account of the development of archeology while interpreting both old and recently discovered material from a novel point of view and daring to progress beyond a classification of stone utensils and other artifacts to conjectures about the ideology of Stone Age people. The book also contains a comprehensive discussion of the problems that Stone Age people solved and were forced to solve; see section [5] [*chapter 1.3*] of the present book.

1.1. Stone Age Art and Knowledge of Nature

[2] Evidence for the foregoing thesis in the domain of *art* can be seen in the astonishing rock images of the late Paleolithic era, which combine in a unique manner technical skills, keen observation, sense of style, desire for novel forms of expression, and ability to quickly realize novel ideas, to form a new whole. They display an economical form of representation, lovely details, and a mastery of isolated phenomena of perspective, alongside a magnificent wealth of colors and impressionistic features. The idea that naturalism is a late form of art that is always preceded by an archaic-infantile stage is thus defeated with one blow. Quite the contrary: the early stages are livelier and follow one another in quick turns (Lascaux), while the later ones, though better painted, are static and lifeless (Altamira). Overall it seems that conventionalism and schematization are later phenomena preceded by periods of stronger creativity. This goes for art as well as science (von Neumann).[10]

Paleolithic art has twice revolutionized a long predominant image of Ice Age humans, which was based on a naïve belief in the gradual, linear mental progress of humans from ultra-primitive "savage" – including contemporary "primitive peoples" as anachronistic relics of the past – to highly sophisticated "civilized peoples."[11] The first revolution consisted in its very *discovery* toward the end of the nineteenth century as well as in the official recognition of its authentic character by the professional world at the beginning of the twentieth

1. Mammoth, Font-de-Gaume, Phase D (partial image)

Drawing by Feyerabend based on Burkitt, Miles Crawford (1963): *The Old Stone Age: A Study of Paleolithic Times*, New York, p. 204.

[10] [*Editors' note: This reference is to John von Neumann. In response to a question by Helmut Spinner, Feyerabend remarked that a more specific reference was not necessary, since he was mentioning von Neumann only as a "daunting example."*]

[11] On Paleolithic art, see Hauser (1951: ch. 1), Breuil (1952), Graziosi (1960), and Leroi-Gourhan (1967) – with excellent color photographs. Marshack (1971) offers a new method of analysis as well as new conjectures about the role of Paleolithic art. Ch. XI of Burkitt (1963) contains a brief and accurate overview.

century (with the breakthrough of the discovery of Les Combarelles near Les Eyzies in the Dordogne in autumn 1903 by Abbé Breuil – a breakthrough famously contested by Breuil's most ardent opponent, Émile Cartailhac).[12] The second was a revolution in the *interpretation* of so-called prehistoric art, which was initiated after World War II by Leroi-Gourhan's work in particular, owing to new material and methods. Other researchers to be mentioned in connection with this revolution, besides Leroi-Gourhan and Marshack, are Annette Laming-Emperaire and Marie E. P. König.[13]

2. Deer, Magdalen, period 5 (art mobilier)

Drawing by Feyerabend based on Burkitt, Miles Crawford (1963): *The Old Stone Age: A Study of Paleolithic Times*, New York, cover image.

[3] We can obtain an approximate notion of those artists' knowledge of nature by drawing analogies to present-day preliterate tribes. A single informant of the Gabon tribe in equatorial Africa offered visiting researchers 8,000 botanical terms. The Hanunoo botanical vocabulary exceeds 2,000, their classifications are sometimes more precise and systematic than those of Western science, and their descriptions of species, subspecies, and variations are sufficiently objective to allow for identification in almost all cases.

---

[12] Cartailhac (1902); on the history of glacial art and its study, see Kühn (1965).

[13] For Laming-Emperaire, see her books (1959, 1962). Leroi-Gourhan summarized his approach in one popular paper (Leroi-Gourhan 1968). Meanwhile Marshack, too, has continued his studies beyond the scope of his aforementioned book; see his papers (Marshack 1972a, 1972b) as well as the subsequent debate and Marshack's reply (Marshack 1972a: 461–77); furthermore his popular presentation in Marshack (1975). For Marie König, see her book (1954) and her essay (1966), but especially her most recent book (König 1973). Marginally notable is also the idiosyncratic interpretation in Giedion (1962). Unfortunately we are still waiting for a monograph that would summarize the current state of research. The best current overview is Ucko and Rosenfeld (1967b), though it does not cover the most recent development associated with the names Marshack and König. However, it provides a detailed discussion of the classic interpretations (Breuil, etc.) as well as of Leroi-Gourhan's new positions, including Laming-Emperaire's closely related approach, and a very informative critical analysis thereof based on anthropological and ethnographic research.

10

Misunderstandings and failures that befell Western researchers using the indigenous classifications were due to insufficient knowledge of the relevant classification principles as well as to qualities not accessible to the differently trained eye of Western observers: "[My teacher] just couldn't understand that it was not the words but the plants themselves that were difficult for me," wrote Eleonore Smith Bowen (1954: 19*) about her attempt to learn an African tribe's botanical vocabulary. "Several thousand Coahuila Indians never exhausted the natural resources of a desert region in Southern California, in which today only a handful of white families manage to subsist. They lived in a land of plenty, for in this apparently completely barren territory, they were familiar with no fewer than sixty kinds of edible plants and twenty-eight others of narcotic, stimulant or medicinal properties" (Lévi-Strauss 1962a: 5).[14] The cornfields belonging to the Guatemala Indians are more rigorously selected according to type than those of their Spanish-speaking neighbors.

> Their fields were quite as true to type as had been prize-winning American cornfields in the great corn-show era when the American farmer was paying exquisite attention to such fancy show points as uniformity. This fact was amazing, considering the great variability of Guatemalan maize as a whole, and the fact that corn crosses so easily. A little pollen blown from one field to another will introduce mongrel germ plasm. Only the most finicky selection of seed ears and the pulling out of plants which are off type could keep a variety pure under such conditions. Yet for Mexico and Guatemala and our own Southwest the evidence is clear: Wherever the old Indian cultures have survived most completely the corn is least variable within the variety. [. . .] It is apparently not true [. . .] that the most primitive people have the most variable varieties. Quite the opposite. It is rather those natives most frequently seen by travelers, the

---

* [Translator's note: Feyerabend appears to have taken more liberties than usual in his German translation of this sentence, which is here translated back into English. For there is no such sentence in the original English text, although it fits the general theme and context of the page he refers to, as the following excerpt from that page shows: "These people are farmers: to them plants are as important and familiar as people. I'd never been on a farm and am not even sure which are begonias, dahlias or petunias. Plants, like algebra, have a habit of looking alike and being different, or looking different and being alike, consequently mathematics and botany confuse me. [. . .] I also found myself in a place where every man, woman and child knew literally hundreds of plants. None of them could ever believe that I could not if I only would" (Smith Bowen 1954: 19f.).]

[14] "Every man has eyes and all his senses to perceive that the world is dead, cold and unending," wrote Jung (1912b: 30); he regards the creation of myth as a revival of the world that happened in spite of the facts. Yet the facts of a pessimistic inhabitant of Switzerland are not the same as those of the Coahuila Indians, who do not just dreamily see a large variety of life in the desert but utilize it for their own sustenance.

ones who live along modern highways and near big cities, the ones whose ancient cultures have most completely broken down, who have given rise to the impression that primitive people are careless plant breeders.

(Anderson 1952: 219)

Just like scientific communities, the Guarani tribal councils in Argentina and Paraguay discuss the terms to be used in the vocabulary and decide which combinations of existing word stems best match the nature of a newly discovered species.

Such an eye for detail, accompanied by a memory unencumbered by written language and an upbringing that places great value on memory skills that from a European standpoint can only be considered extraordinary, also explains the knowledge of seafaring peoples such as the Polynesians. They use halos, star colors, variations in star brightness, the appearance of sun, moon, and close star groups such as the Pleiades, cloud shapes, cloud colors, and the curvature of the Milky Way, which changes with the seasons and can thus be regarded as a variable condition, to predict weather, wind direction, and wind force. Even Herder already observed that "[T]he sea-faring folk still remain particularly attached to superstition and the marvellous. Since they have to attend to wind and weather, to small signs and portents, since their fate depends on phenomena of the upper atmosphere, they have good reason to heed such signs, to look upon them with a kind of reverent wonder and to develop, as it were, a science of portents" (Herder 1769b: 71). Enlightened seafarers such as Captain Cook were surprised by the Polynesians' "reverent wonder and development of a science of portents": "Of these rules I shall not pretend to judge; but I know that, by whatever means, they can predict the weather, at least the wind, with much greater certainty than we can" (as quoted in Makemson 1941). They possessed not only precise mastery of the characteristics of trade winds and the statistical distribution of tropical cyclones but also knowledge of ocean currents, of indicators of land still beneath the horizon (cloud formations, the flight of birds, the reflection of green lagoon water in suitably located clouds, volcanoes in the night, etc.).

Many of the empirical-statistical regularities involved are unknown to Western scientists, not easily accessible to their research methods, and also not needed in light of alternative available tools. The astronomy needed on the voyages is taught in special schools and passed on only hesitantly to outsiders. The lessons are in oral language, which means that students' memories first have to be trained in such a way that they will be able to memorize long passages in technical jargon without the aid of writing them down. The poets and narrators of the *Iliad* and the *Odyssey*, as well as Hindu astronomers, who as late as in the nineteenth century were capable of precisely predicting lunar eclipses to within a few minutes by using mental calculations based on memorized rules, have demonstrated that the human brain is capable of such feats. Astronomy classes for Polynesians cover lodestars for great circle paths, reduction to the horizon for various latitudes and for a lodestar height of up to 30 degrees, zenith stars and zenith constellations for various latitudes (which presupposes a vague idea of the nature of a circle of latitude), stars of the same declination capable of consecutively acting as lodestars, knowledge of azimuth changes in lodestars with variable latitude, the study of systematic line networks on the starry sky used to make precise estimates of star heights to a few degrees, weather indicators, and cosmological theories and legends containing the (meteorological, astronomical, etc.) main events along important sea routes in easy-to-memorize verses, thus enabling their repeated use.

The great expeditions of the tenth to fourteenth centuries were attended by specialists, one or two astronomers, prophets, sailing masters, and navigators. In the islands themselves they followed up the routes of sun, moon, and major stars in observatories usually consisting of a slab (horizon) and various guide stones. Given the degree of precision

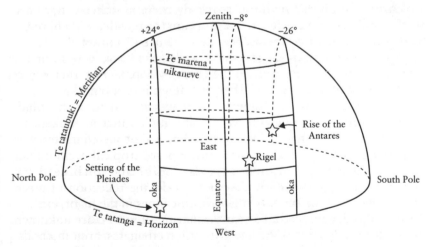

3. Reconstruction of Gilbertese sky dome

Lewis, David (1974): "Voyaging Stars: Aspects of Polynesian and Micronesian Astronomy," in F. R. Hodson (ed.), *The Place of Astronomy in the Ancient World*, London, p. 136.

[*Translator's note: In Gilbertese astronomy, "oka" are imaginary rafters supporting the sky.*]

of the astronomical observation required and achieved for professional navigation, we can expect that these early astronomers did not overlook precession, and indeed, myths about the destruction of "Earth" and "Heaven" as well as of the creation of a new "Earth" and "Heaven" do exist in the Pacific region up to the coast of South America.

The common assumption is that the Polynesians' voyages ceased in the fifteenth century (that is, long before the development of the Pacific region by Western explorers), that their meteorological, astronomical, etc. background knowledge was no longer completely preserved even in the eighteenth century, and that today there is nothing but some deformed remnants left of it. Thus, the complete body of knowledge has to be inferred via an interpretation of the surviving traditions. Maud Makemson (1941 – including a star catalog with 772 entries) chose this route. Her results, which I largely follow in this essay, have been recently challenged in several respects. (1) The interpreted texts allow for alternative interpretations and are in part contradictory. (2) The stated astronomic methods fail for the distances alleged. (3) The traditions themselves are not reliably authentic. Andrew Sharp, one of the strongest critics of the theory that the Polynesians conducted repeated journeys over great distances, writes about New Zealand traditions (according to William Colenzo): "In all this mythical rhapsody there is scarcely a grain of truth; and yet some educated Europeans have wholly believed it. The New Zealanders themselves never did so" (Sharp 1963: 79). (4) Archeological finds suggest that many of the voyages went in only one direction. (5) Furthermore, the voyages were so rare that we can hardly speak of degeneration after the fifteenth century.

The debate has yet to come to a conclusion. Some of the objections rest on mere considerations of plausibility, that is, on assumptions about the degree of intelligence that a Polynesian needs when employing methods that are in principle practical. "It is unlikely that the navigational method used was LATITUDE SAILING (parallel sailing)," wrote Kjell Åkerblom (1968: 47),[15] even though David Lewis, while on a voyage without equipment from Tahiti to New Zealand, demonstrated that the margin of error does not exceed 26 nautical miles.[16] In 1969 the Carolinian navigator Hipour crossed the open sea three times over a distance of 720km, relying solely on astronomical observations made with the naked eye as well as on

---

[15] A book that otherwise offers a good and relatively complete overview of various works featured in the debate.

[16] See his reports (Lewis 1966a, 1966b), as well as his popular account (1974b) and the essay by Emory (1974) that preceded the latter.

oral traditions of Carolinian seafarers that had not been used in three generations.[17] Facts such as the existence of star myths with northern connotations in southern latitudes indicate the pan-oceanic distribution of astronomical ideas used by the seafarers.[18] The use of computers introduced a new aspect into this debate, because it showed that a mere drifting of the vessels cannot explain the numerous crossings of the Pacific that actually took place. For that it was necessary to sail consistently before the wind, which means that the seafarers crossed the ocean purposefully, not randomly.[19] The mythic form (ballads, verses, stories with heroes and minor characters) is an essential support for a preliterate science and poetry and is for this reason quite common. It plays a part similar to that of the binary system today but is more adaptable and useful in securing social cohesion.

1.2. Megalithic Astronomy (Stonehenge)

[4] *Stone observatories*, the results and predictions of which are preserved only in the memory of preliterate astronomers and priests and in the structure of the buildings, can be traced back to even much earlier times. Stonehenge in England is an impressive example.[20] Three different waves of peoples worked on this monument between the years 1900 and 1600 BC, each leaving after completion of their work. The first wave consisted of late Stone Age hunters and farmers from the European continent. They dug the exterior circular groove, a circular row of holes (the so-called Aubrey holes), and erected the upright standing boulders – among them the so-called "Heel Stone" – 30 meters outside of the groove. Today this phase is

---

[17] Lewis (1971) and Goodenough (1953).

[18] Lewis (1974a). On the training of Polynesian seafarers, see Best (1923); on their astronomical knowledge, see Best (1922); and on the debate about Sharp's theses, see Golson (1972).

[19] Levinson, Ward, Webb, et al. (1973). On the memory skills of preliterate astronomers, see Neugebauer (1952). On the esoteric aspect, see Lewis (1974a: 136ff.). It was not until the year 1969, when a large portion of their science was already lost due to the impact of Western intruders, that the leaders of the Tu'i Tonga's clan agreed to disclose to the author the remainder of their more esoteric concepts.

[20] Stonehenge is described and discussed in Hawkins (1965), from where the drawings and photographs of Stonehenge in this book were taken. Hawkins first introduced the scientific world to his ideas in his sensational Stonehenge articles "Stonehenge Decoded" and "Stonehenge: A Neolithic Computer" in *Nature* (1963/4). On the debate about Stonehenge, see Atkinson's critical review of Hawkins' aforementioned book in Atkinson (1966), as well as Hawkes (1967), Hawkins (1968), and Hoyle (1966), with commentaries by Hawkins, Atkinson, Thom, and others in *Antiquity*, 41 (1967), pp. 91–8.

called Stonehenge I. Its builders' successors were first the Bell-Beaker people (Stonehenge II) around 1750 BC and then, around 1700 BC, the first tribes of the Bronze Age. The latter erected Stonehenge III and used the basic plan of Stonehenge I and II for their purposes. This is noteworthy: people of different cultures and different races recognized the astronomical significance of the monument and improved it by building new additions – a first indicator that there was an *international astronomy* in Europe during the late Stone Age the elements of which were known beyond tribal borders. In the end Stonehenge was spread out over 140 meters. Each of the upright standing boulders in the circle of Stonehenge III (diameter *c.*30 meters) weighs *c.*25 tons, and each of those in the horseshoe-shaped center area between 40 and 50 tons. The weight of the circle's capstones is about 7 tons, and the weight of the smaller inner circle's dolerite boulders about 5 tons each.

The building material was hauled over great distances; the dolerite boulders over 390 kilometers on water and on land, the larger sandstone boulders over roughly 32 kilometers. The workmanship that went into the stones (note the convex curve applied to the vertical stones!) shows great finesse and suggests (in combination with other indicators) Mycenaean influence. It has been estimated that the number of minimum man-days to complete the entire monument was 1.5 million. The organizational problem was increased by the fact that the building site was flooded with peoples of different cultures – and yet the basic architectural plan remained the same even in its details over a period of 300 years, and additions to the monument were built in strict consistency with that plan.

We should *assume* that the building's function was at least in part mythical or religious. We *know* that astronomical observations and predictions played an important part: circle and horseshoe convey an impression of symmetry and regularity. The minimum distance between the adjacent upright boulders of the horseshoe is 30 centimeters, and the circle's boulders are all placed at the same distance with a maximum error range of 10 centimeters. From inside you can look through the rifts between the boulders in only seven directions; everywhere else the view either is blocked by boulders or goes through places where we know that boulders used to stand. The directions emphasized correspond to the declinations ±24, ±29, and ±19. Sixteen noteworthy directions of an earlier building period (Stonehenge I) as well as the direction from the center to the "Heel Stone," a huge single boulder outside of the ring-shaped groove

4. Reconstruction of Stonehenge II

From: Singer, Charles, E. J. Holmyard, and A. R. Hall (1967): *A History of Technology. Vol. 1: From Early Times to Fall of Ancient Empires*, Oxford, p. 492, fig. 321.

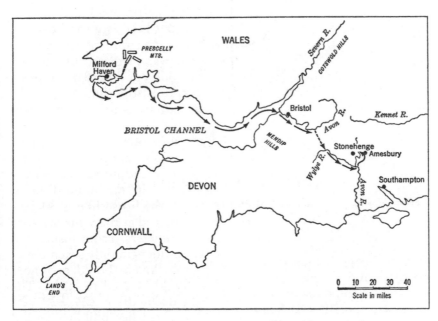

5. Probable transportation route of the "bluestones"

From: Hawkins, Gerald S. (1965): *Stonehenge Decoded*, Garden City, p. 64, fig. 6.

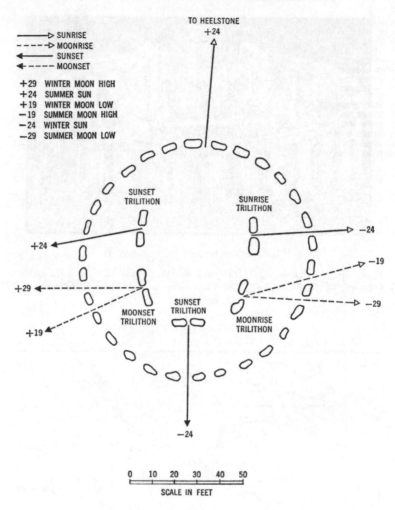

6. Boulder arrangement of Stonehenge

From: Hawkins, Gerald S. (1965): *Stonehenge Decoded*, Garden City, p. 108, fig. 12.

of Stonehenge I, indicate the same declinations, and they are the northern and southern maximum of the sun's orbit and two northern and southern maximums of the moon's orbit, so that moonrise, moonset, sunrise, and sunset can be observed in midwinter and midsummer with an average error margin of (vertically) only 1.2 degrees.

18

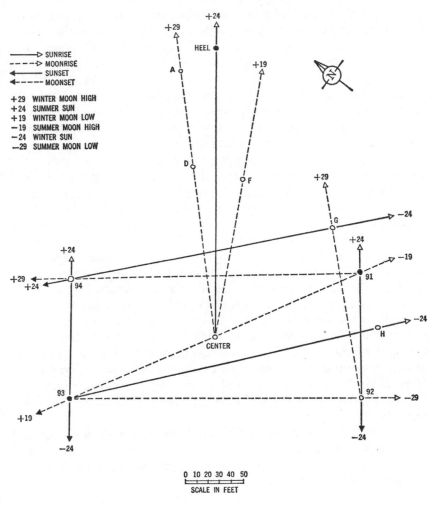

7. Symmetrical structure of Stonehenge

From: Hawkins, Gerald S. (1965): *Stonehenge Decoded*, Garden City, p. 107, fig. 11.

Such recording of asymmetrical events by means of symmetrical structures such as circles and horseshoes, which at first glance have nothing to do with reality but do enable the expert to gain insights into astronomical processes, is an early predecessor of the contemporary attempt to gain knowledge of nature based on principles of symmetry. Symmetries are present already in Stonehenge I in the form

19

of a rectangle whose sides are the directions of the sun and moon maximums and whose diagonal is the direction of the smaller moon maximum. Only in one location in the northern hemisphere does the arrangement correspond to actual astronomical conditions, and Stonehenge is practically at that very location (even in Oxford the deviation from the right angle is clearly noticeable). It can therefore be concluded that the observatory was built on the basis not only of local observations but also of knowledge of the geographic variations of astronomical phenomena. Furthermore, Stonehenge appears to have been used to predict eclipses; an eclipse occurs whenever the winter moon rises above the Heel Stone. Thus, this structure functions as heavenly clock, forecast station, and religious arena all in one; we should never underestimate the propagandistic skills of the guardians of knowledge, who are always somewhat influenced by religion. Their success serves to strengthen both faith (knowledge) and the power of the priests. This can be demonstrated with a few concrete cases. For example, the light of the sun illuminated the statue of the god in the Holy Chamber at the Amun-Ra temple in Karnak, which was hidden in the dark at the end of a hallway, just at the time of the summer solstice, thus announcing the prompt, rapid rise of the Nile and reinforcing faith in the power of the priests. Changes in the obliquity of the ecliptic compelled the priests to gradually rebuild the optical path (reconstruction in Lockyer 1964: 104). *Today* we produce science shows in order to impress sponsors, including naïve taxpayers, and to tempt them into funding expensive projects.

Atkinson (1960) contains a detailed description of Stonehenge and its historical background that rejects the astronomical "moonshine" interpretation. Hoyle summarizes the arguments in favor of the opposite position as follows:

> It is *not* a speculation to assert that *we* ourselves can use Stonehenge I to make eclipse predictions. We could certainly do so without making any substantive changes in the layout. While this does not *prove* that stoneage man did in fact use Stonehenge I for making eclipse predictions, the measure of coincidence otherwise implied would be quite fantastic. How does one *prove* any incident belonging to the past? Historians argue by documentary evidence. But how if their documents are false? A plethora of documents belonging to the present day are false, many of them made so deliberately. It is not possible to argue that Stonehenge I was falsified deliberately, to maintain a façade of astronomical subtlety by a people ignorant of astronomy. It is probably hard for the historians to accept the idea of a geometrical arrangement

of stones and holes providing evidence much stronger than a document; but I believe this to be true.

(Hoyle 1972: 51*)

Today this view is shared by all experts including Atkinson, though the latter points out that the model of a Stone Age astronomy in Europe "do[es] not fit the conceptual model of the prehistory of Europe which has been current during the whole of the present century" (Atkinson 1975: 51). Alexander Thom appears to have been the first to point out that the megalithic astronomers used distinctive features of distant hills as reference points in long lines of sight.[21] If we aim at the sun against the side of a hill parallel to the daily orbit, we can determine the slight alterations of the declination around the solar eclipses (roughly 12″ per day) and thus the length of a year with relative precision: at a distance of 30 kilometers a 12″ angle corresponds to a lateral shift of 1.80 meters. An aiming point has to be placed on this location in order to still see the setting sun flash up on a suitable slope of the hill. We can find distinctive hill formations with matching aiming points in numerous locations in England and Scotland. Moon maximums and minimums were extrapolated from three observations, and in this way a value for the obliquity of the moon's orbit was obtained that differs only by 9″ from the value we can get today. The mathematics used in this process, which was developed and preserved solely in the priest astronomers' memory, included knowledge of the Pythagorean theorem.[22] Now that radio-carbon dating has been corrected on the basis of results obtained by Suess and others (Renfrew 1971, 1973), it is no longer so easy to trace this knowledge back to oriental or Mycenaean influence; rather, it should be regarded as the result of an independent development.

---

* [Translator's note: Feyerabend himself did not include a page number in his reference. Nor did he include the emphases in italics that were part of the original English quote, which he had translated into German in his typescript. As in some of his other translations from English into German, he took some freedoms that slightly altered the meanings; for example, his German translation reads "Stonehenge" instead of "Stonehenge I."]

[21] Thom (1966, 1969) contain a discussion of megalithic observatories in consideration of long lines of sight; Thom (1967, 1971) include summary reports. The most recent contribution by Thom, his son, and his grandson is Thom, Thom, and Thom (1975).

[22] The ideology of the priest astronomers, who used the observation points, and their relation to the later druids are discussed in MacKie (1974), which also contains a comprehensive analysis of Scottish observatories. Meanwhile, the sciences needed to examine such old cultural assets have united to form a new scientific discipline, so-called astro-archeology or archeo-astronomy. Aveni (1975) offers a new overview of this branch of science with a special focus on astronomy in pre-Columbian America.

[5] For a long time the learned world greeted these results with astonishment, disbelief, and ironic criticism.[23]

> It is important that non-archeologists should understand how disturbing [. . .] are [these implications], because they do not fit the conceptual model of the prehistory of Europe which has been current during the whole of the present century, and even now is only beginning to crumble at the edges. Part of the foundations of this model can be summed up in the phrase *ex oriente lux* – the idea that cultural, scientific and technological innovations were made in the early civilizations of the ancient east, and reached Europe only in a diluted and etiolated form through a slow and gradual process of diffusion. In terms of this model, therefore, it is almost inconceivable that mere barbarians on the remote northwestern fringes of the continent should display a knowledge of mathematics and its applications hardly inferior, if at all, to that of Egypt at about the same date, or that of Mesopotamia considerably later.
>
> (Atkinson 1975: 51)

1.3. Critique of Primitivist Interpretations of the Prehistoric Era

The view of the barbaric nature of the megalithic astronomers is due not least to the method used by archeologists. This method is inductive and consists of the following steps: first we collect utensils, works of art, and other evidence; then we classify them according to similarities pertaining to form, material, processing of material, topics, and so forth; then we merge complexes of similar cultural traces as separate "cultures," name them according to a main location of finds and study their dispersal; we try to determine each culture's age on the basis of the age of the geological layer in which the evidence was found, and we attempt furthermore to locate the "carriers" of the culture in question by studying remnants of skeletons. None of these steps leads to a theory, and the newly introduced culture concept is simply a collective name to represent the discovered material. The next step consists in an "interpretation" of the collected material. At

---

[23] [*Editors' note: In an earlier version section [5] started right here. However, due to extensive edits of sections [5] and [6] Feyerabend completely obliterated the separation of [4] and [5]. Furthermore, the two versions of the typescript do not entirely conform with regard to sections [5] and [6]. The version from the Feyerabend archive is notably more comprehensive in these places than the version available to Helmut Spinner. We have followed throughout the more comprehensive last manuscript version. The quotations by Atkinson and Aveni that were used show that Feyerabend must have undertaken his final edits after 1975.*]

this point the procedure changes radically. While the archeological work is diligent and precise, rarely going beyond what can be determined with certainty, many interpretations are content with vague generalizations. Images of animals have a "magical" meaning and are elements of "hunter magic." Depictions of pregnant animals or women play a part in "fertility rituals." Oblong or roundish objects that cannot be further specified have a "sexual meaning"; mixed images such as horses with bear paws are "naturally" religious and so forth. It should be noted that these interpretations are based not on a detailed examination of the available evidence, but on general notions regarding the "evolution" of humans, which occasionally are enriched by vague ideas stemming from the study of surviving Stone Age cultures. The notion, however, that early humans were "tool makers," which we find in nearly all accounts and which by now has turned into a basic research postulate itself, assumes that the Stone Age humans' domain of activity and thought is fully determined by the body of cultural traces that they left behind.

How wrong this assumption is becomes clear when we apply it to the archaic Greek tribes on the counterfactual premise that all oral traditions – including the *Iliad* and the *Odyssey* – have been lost. Furthermore, we can easily see that much of the knowledge previously discussed does not leave any material traces: the circle of notions entertained by a group of humans is much more comprehensive that what can be inferred from its cultural remains. And there is a way in which we can gain access to those notions. We know now that *Sinanthropus pekinensis* lived about 500,000 years ago in Zhoukoudian and used fire. Let us now consider what kind of knowledge a fire-using culture requires.

We must assume, then, that the brain capable of maintaining a fire culture would be capable of learning that wet wood would not burn as well as dry, that most spring or summer wood with the sap flowing thick would probably not burn as quickly or as well as dead wood or winter wood, that scraping or cutting a branch in the spring could foul or gum the hand, that green grass or reeds do not burn as well as yellow or brown, that summer leafing wood smoulders, that the wood you wish to use that day must be kept out of the rain, that one does not stand close and downwind of heavy smoke, that in the cold of winter fire serves most, that in the night it offers light and warmth and safety since most wild animals fear fire. [. . .] Far more symbolically – and again, with or without a large use of language – fire is "alive." It must be tended; it needs a home and place out of the great winds, the heavy rains, the deep snows; it must be constantly fed; it sleeps in embers and can die, yet it can also be blown back

23

to life by the breath; it can burn a hand; it sputters angrily and brightly with animal fat; it dies entirely in water; it whispers, hisses, or crackles, and therefore has a variable "voice"; it uses itself up, transforming a large weight of wood into gray ash, while climbing by smoke and savor to the sky, at last disappearing in the wind; one can carry its spirit or "life" on a burning branch or ember to make a second fire. To a man with fire, then, there is a continuous involvement in a complex, dynamic process which creates its varying, yet "artificial," demands, relations, comparisons, recognitions, and images. More important, perhaps, fire ties one down in time and thought because of the constant requirements in maintaining it. These demands may have been greater at the beginnings of fire culture, particularly if early man used fire without an adequate skill in sparking or making it. In the old anthropological and archaeological concept, fire "freed" man to live in new climates and lands. But clearly, it also *bound* him strictly, culturally and functionally. Even if it freed the hunter temporarily, while the women and children were rooted safe around a fire, the group was bound in time and place.

(Marshack 1971: 112f.)

What goes for fire also goes for other processes. There were, for example, the regularities of pregnancy and menstruation, as well as the annual seasons and the changes in nature associated with them, such as migrations and biological metamorphoses of fish, bison, mammoth, lion, ibex, and reindeer. There were also the various species-specific reactions of hunted prey and dangerous predators as well as individual variations among them. There were the differences in human character, diseases, death, the phenomenon of birth. There were the characteristics of the material used to manufacture weapons, works of art, and dwellings, material that interacted in complex ways with the talents of the artisans. There were the characteristics of the moon's orbit, the gradual increase, the full moon, the gradual "dieback" of the moon as well as a brief period of almost deathlike disappearance, which is soon followed by a new rebirth. The tides on the coast were to be noted, sometimes greater, sometimes smaller, depending on the position of sun and moon. Many of these regularities are integrated into the cycle of an animal's life and thus did not have to be relearned by humans. The human innovation consists in an addition of "artificial" processes such as fire and in an augmentation of human knowledge beyond the naturally obtained quantum. All of these factors need to be taken into account in our interpretation of remaining cultural traces.

In the interpretation of the evidence provided by archeological inves- tigation it is important to realize its limitations and to appreciate the complexity of the factors involved. [. . .] It will be apparent that

24

an onlooker, seeing [contemporary Stone Age-type people, e.g. in Australia] at different seasons of the year, would find them engaged in occupations so diverse, and with weapons and utensils differing so much in character, that if he were unaware of the seasonal influence on food supply, and consequently on occupation, he would be led to conclude that they were different groups. [. . .] It can be shown that little would remain to suggest to an archeologist of the future [. . .], forced to depend upon the examination of old camp sites and such artefacts as resisted decay, the extent and the complexity of the culture. [. . .]

Within the bounds [. . .] of a single [. . .] territory a people may spend several months a year as nomadic hunters, in pursuit of bush game, wild honey and small mammals, and exploiting the resources of vegetable foods [. . .] A few months later the same people may be found established on the sea coast in camps that have all the appearance of permanence or at least of semi-permanence, having apparently abandoned their nomadic habits. They will remain in these camps for months on end, engaged now in fishing and in the harpooning of dugong and turtle from canoes; leading, in fact, the life of a typical fishing and seafaring culture. In each case the camps and the house types, the weapons and the utensils, are of a specialised type and related to the seasonal life [. . .] [The] relatively rich material culture [. . .] would yet leave only the slenderest of evidence for later archeological investigation.

The seasonal factor is recognized by the aborigines themselves, and stressed by the fact that they have classified the types of country, as accurately and as scientifically as any ecologist, given to each a name, and associating it with specific resources, with its animal and vegetable foods, and its technological products. [. . .] [I]n spite of its complexity, this sophistication, which is not generally credited to people of a "stone age" culture, little enough would remain on account of the ephemeral nature of the greater part of the material used, to provide a key to the true state of affairs or even to correlate as part of a seasonal cycle the occupational sites which are visited regularly at the appropriate season. Too often the nomadic movements of such people have been regarded as merely aimless or random wanderings; studied at first hand they assume a very different character. [. . .] [T]hey form a regular and *orderly* annual cycle carried out systematically, and with a rhythm parallel to, and in step with, the seasonal changes themselves.

(Thomson 1939: 209, 211)

The attempt to obtain an understanding of Stone Age culture solely inductively from cultural traces left behind as evidence by Stone Age peoples, that is, without taking into account additional presumptions about their environment, knowledge, mentality, and problems is doomed to fail. Thus let us assume that Stone Age humans are not different from us (they are, after all, fully developed examples of the

25

species *Homo sapiens*), that they possessed curiosity and intelligence comparable to ours, and let us not forget that Stone Age economy did not demand more than four hours of labor per day from people.[24] Thus we obtain the following picture.

1.4. The Dynamic Worldview of Stone Age Humans

The knowledge that we have described thus far, and that is needed for managing a life with the utensils and circumstances familiar to us, is knowledge of *temporal processes*, not of rigid entities. The temporal processes recur more or less in the same manner and establish stable entities due to their manner of repetition and their overlap with other temporal processes. The temporal form is primary, while non-temporal entities either do not occur at all or are derived from the temporal forms. The more detailed our basic knowledge of nature becomes, the more "artificial" knowledge – such as the knowledge of fire – we acquire, and the more we enter into a situation in which merely reacting to our environment no longer suffices. In addition to both acquired and genetically determined *reactions* to the environment there is a new form of interaction through *traditions*. On the basis of the previous considerations we are now able to assign these traditions certain general properties: they need to possess means of expression in order to represent sequences of events, that is, *they have to contain reproducible processes of a certain kind*. Of course, the simplest kind of representation of a sequence of events is an *example*: we illustrate how to create and maintain a fire by lighting one, letting it burn, protecting it from wind and rain (or simulated wind and rain), and rekindling it when it is about to die. As we increase the element of simulation we obtain a process similar to a ritual, such as a ritual simulation of birth, of growth, of blossoms, of the moon's waning. This ritual does not initially have anything to do with magic: we do not aim to get the *moon* itself under control but rather our *knowledge of the moon*, and we wish to represent this knowledge in a way that better enables us to *see* the moon *correctly*. At a more abstract level and with an already further developed language, the intellectual (oral)

---

[24] On the economy of Stone Age hunter societies, see L. K. Binford and S. R. Binford: "There is abundant data which suggests not only that hunter-gatherers have adequate supplies of food but also that they enjoy quantities of leisure-time, much more in fact [. . .] than professors of archeology" (Binford 1969: 328). (Do the myths of a Golden Age refer to this state?) Thus the introduction of agriculture was not a necessity, at least not everywhere.

26

act replaces the ritual: we then have a *story*, a *myth* that describes identifiable events in the moon's orbit by means of identifiable story sections. As moon, sun, people, and animals are brought into relation to one another, the complexity of such stories can increase considerably, and yet the basic structure remains the same: it is a means to represent sequences of episodes, not of objects.

This property of Stone Age representational means has to be encountered in the cultural evidence on which science has hitherto based its interpretations; this is Alexander Marshack's basic thesis, with which he revolutionized our interpretation of Paleolithic works of art and utensils. It can be expected that such evidence contains depictions of processes that have previously gone unnoticed and that can be associated with quasi-periodic natural processes, such as the movement of the moon. Then depictions of animals and hitherto unidentifiable symbols may possibly represent situations at certain seasons of the year, that is, coincidences or temporally determined evolutionary sequences. Research has confirmed all these presumptions, though only after diligent, microscopic examination of the hitherto unidentified details. This examination identified many "sexual symbols" as plants that are in blossom at the time of the depicted animals' rutting season, and it also showed that many animals were depicted in various states of pregnancy or maturity. Abstract symbols, however, could be coordinated with the moon's procession in most surprising ways. According to Marshack,

> The repertoire of images found across Upper Paleolithic Europe, therefore, suggests a storied, mythological, time-factored, seasonal, ceremonial and ritual use of animal, fish, bird, plant, and serpent images, and it apparently also includes at times what seem to have been selective and seasonal killing and sacrifice, either of the image, in rite, or of the real animal. The complexity and interrelation of these storied meanings cannot easily be explained by any generalizing theories propounding concepts of hunting, magic, fertility ritual, or sexual symbolism. Instead, the art and symbol suggest a broad range of cognitions, cultural and practical, and a profound understanding of processes in nature and of the varieties of living creatures.
>
> (Marshack 1971: 260f.).

Animals and animal combinations (such as ibex and fish, which recall the constellations and likely indicate their forerunners) refer to annual seasons; complex processes in nature are structured, stipulated, and passed on by means of simple stories. These pictorial stories are supplemented by "a common, basic tradition of notation in all the

27

European Upper Paleolithic cultures, and this notation was cumulative, 'time-factored,' and possibly lunar" (Marshack 1971: 108). This Paleolithic informal notational system, which was first tentatively decoded (figures 8, 9, 10) in a Mesolithic bone – the so-called Ishango bone – can be found in much older material and is again identifiable in full detail only through a microscope.

We can only assume that the thought instruments outlined here and passed on over many generations were still used at the time of the first written record of the basic stories containing them. Though the original "pure form" will likely be disturbed and covered by older ideas and outgrowths, attempting a reconstruction is not entirely hopeless. Following older interpretations, Georgio de Santillana (de Santillana and von Dechend 1969) attempted to understand myths from this point of view, and he indeed discovered in them traces of an international, chronologically structured astronomy composed of *stories*. After the revision of the radiocarbon dating method, neither the distribution of these ideas nor the spread of agriculture, certain types of graves, or architectural principles can be explained solely in terms of diffusion: the same brain reacts to similar problems with similar solutions. Thus archeology, mythology, philosophy, and anthropological speculation approach their subject using different methods, yet in the end they come to the same findings, namely that archaic humans were more advanced and closer to the present age than science and lay opinion have been willing to admit for a long time.

The progressiveness of Stone Age people and of the mythologists following their traces becomes clearer when we compare their ideas with those of later philosophers and scientists. We saw that Stone Age people *participated* in numerous natural processes and *reflected* them in their traditions through media that possessed a process character themselves. A dynamic world is dynamically represented. Even for Hesiod the world was a gigantic process composed of many short sequences of events, and rest is only apparent, since it is but the result of tendencies keeping each other temporarily in balance. Parmenides initiated the attempt to represent motion by means of stable principles generally incapable of change, and that attempt promptly failed: motion did not get merely misrepresented so much as excluded from the representation of nature as altogether non-existent. This rigidity ("eternal laws of nature") lasted into the nineteenth century, expressing itself also in the fact that the laws of nature did not refer to the real world but rather to an ideal world of powerless motion and closed systems, until finally experiments such as Prigogine's in thermodynamics reintroduced the older point of view. Thus "most

28

8. Images of the Ishango bone from the Kongo: Mesolithic

From: Marshack, Alexander (1971): *The Roots of Civilization: The Cognitive Beginnings of Man's First Art, Symbol, and Notation*, New York, p. 23, fig. 1a, b, c.

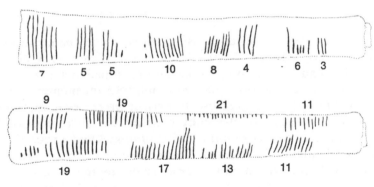

9. Drawing of the markings on the bone

From: Marshack, Alexander (1971): *The Roots of Civilization: The Cognitive Beginnings of Man's First Art, Symbol, and Notation*, New York, p. 23, fig. 2a, b.

recent" science is a combination of both "modern" science and the ideas of Stone Age philosophers and scientists.

[6] The reason for the professional world's reluctance to accept these insights was "because they do not understand them, or [. . .] because it is more comfortable to do so" (Atkinson 1975: 51): it was *more comfortable* to stick with the interpretations that had been hitherto used to explain the material, and that were based on vague

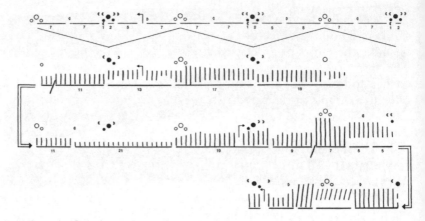

10. The significance of the lines on the bone in relation to the phases of the moon

From: Marshack, Alexander (1971): *The Roots of Civilization: The Cognitive Beginnings of Man's First Art, Symbol, and Notation*, New York, p. 30, fig. 3.

and not even very imaginative assumptions about the evolution of humans. Thus, a busty "Venus" such as the Venus of Willendorf "naturally" had to be a sex symbol, realistic animal images were used for hunter magic, line drawings with multiple appendages – which today are recognized as representations of processes – display the "primitive" character of the culture examined. Often it is simply *ignorance* that obstructs genuine understanding. On occasion this has been openly admitted. Augustin Krämer, for example, writes about the course of his research on Polynesian navigation:

> My old friend Le'iato of Tutuila tried very hard to explain to me that kind of navigation. Once when he had come to me for an evening for that purpose and was unable to complete his lecture because some stars were missing, he woke me up a few hours later in the middle of the night and although I had meanwhile attended a festivity in Apia he did not let me go until the sun rose. I must admit I did not grasp everything, [. . .] for I am not much of an astronomer.
>
> (Krämer 1903b: 283)

Archeologists, anthropologists, even classic philologists rarely have expertise in astronomy, and yet they usually don't admit that their knowledge is inferior to that of the Stone Age people.

Ignorance very frequently hides behind academic jargon. This goes especially for psychoanalysis, which is a blessing for all those thinkers who love to lecture about "human nature" without having to make the effort of a detailed study. Freud, for example, wrote in *The Interpretation of Dreams*: "The dream, which in fulfilling its wishes follows the short regressive path, thereby preserves for us only an example of the primary form of the psychic apparatus which has been abandoned as inexpedient. What once ruled in the waking state when the psychic life was still young and unfit seems to have been banished into the sleeping state" (Freud 1900b: 447). And Jung commented: "All this experience suggests to us that we draw a parallel between the phantastical, mythological thinking of antiquity and the similar thinking of children, between the lower human races and dreams. [. . .] Consequently, it would be true, as well, that the state of infantile thinking in the child's psychic life, as well as in dreams, is nothing but a re-echo of the prehistoric and the ancient" (Jung 1912b: 19f.). The parallel fails due to the detailed expertise that we find in "the prehistoric and the ancient" as well as with the "lower human races," and which has nothing "infantile" or "dreamlike" about it. On the contrary, "infantile and dreamlike" is how a way of thinking such as Parmenides' appears to us now, since it denied the existence of motion, thereby leading Western thinking astray for centuries. "Infantile and dreamlike" is also how we must view Jung's theory, which, in spite of the knowledge available in 1912 and of the simplest considerations of plausibility (such as, should we assume that the inventors of fire, agriculture, literature, stonemasonry, and animal breeding came across all of this only in dreams and in infantile play?), regresses in a dreamlike manner to earlier knowledge periods, thereby creating a "restorative for the brain, which during the day is called upon to meet the severe [?] demands for trained thought" (Jung 1912b: 22). Admittedly, a purely naturalistic explanation of myths – that is, an explanation that aims at deriving myths solely from facts of nature such as the sun's orbit, the moon's motion, planetary motions, constellations, etc. – leaves unanswered the question why the sun's progression is imitated by the images used in the myth (Rank 1909b: Introduction). For even the most precise scientific theories are never completely determined by the facts of nature. And yet the outline of an explanation offered by Jung is far too simplistic.

David R. Dicks (1966) is one example of expert bias. Dicks wrote: "Conditioned as we are by over 300 years of scientific discovery [. . .] we find it very difficult to understand a world where science played in effect a very insignificant part" (Dicks 1966: 29). This is true, and it applies especially to Dicks himself. He says that the determination of the equinoxes "presupposes a spherical earth as the central point of a celestial sphere with equator, tropics, and the ecliptic [. . .] inclined at an angle to the equator [. . .] and, of course, it presupposes a knowledge of the length of the solar year and a fixed calendrical scheme" (Dicks 1966: 32f.). The inhabitants of the Gilbert Islands manage without this knowledge (although they possess some of it). They determine azimuths in terms of the direction from a slab, which serves as observation point, to neighboring islands. The rising azimuth of the equinoctial sun is named *Bike-ni-kaitara*, which roughly means "island of standing face-to-face." But what is standing face-to-face here? It is the island of the rising azimuth and the island of the setting azimuth. Is it probable that the Polynesians discovered such relations? It is not only probable, but quite certain. Their method of navigation consisted in selecting two diametrically opposed horizon stars as lodestars and trying to keep the boat in the direction they indicated. What would be more natural, then, than to be interested also in the point opposite to a certain rising azimuth of the sun? The Greeks, who were Dicks' subject of study, were coastal navigators and probably not familiar with such methods. They were, however, familiar with the *gnomon*, which enabled them to determine the equinoxes in the following manner: they observed when the rising and the setting shadow constitute a line. The days on which this happens are the days of the equinoxes.[25]

[7] The insight that "primitives" of the present and the early past (Stone Age) had considerable *expert knowledge*, that they were able to formulate this knowledge theoretically and to utilize it practically, forces us to pay the fruits of their *thinking* more respect than has been done up to now. We cannot rule out the possibility that their myths are at least in part a spin-off of their factual knowledge and constitute an early phase of our own knowledge about nature. Indeed, we should be ready for the discovery that this early phase contained elements more accurate about nature than the conjectures and methods of contemporary science. In this respect Aristotle was far more realistic and farsighted than his most recent successors. His introduction to philosophy begins with the oriental magicians (Aristotle, *Fragments* 6); he continues his account with the "theologians" and opticians, discusses the sayings of the Seven Sages, and lays the foundation for a collection of proverbs and empirical instructions,

---

[25] For details about the inhabitants of the Gilbert Islands, see Grimble (1931).

which he conceives of as remnants of a preliterate philosophy – an assumption that modern thinkers should adopt as well for the sake of completeness. Undoubtedly this conclusion was based on his doctrine that in history the same truth occurs not only once but frequently and in various disguises:

> Our forefathers in the most remote ages have handed down to their posterity a tradition, in the form of a myth, that these bodies are gods, and that the divine encloses the whole of nature. The rest of the tradition has been added later in mythical form with a view to the persuasion of the multitude and to its legal and utilitarian expediency; they say these gods are in the form of men or like some of the other animals [. . .] while probably each art and each science has often been developed as far as possible and has again perished, these opinions, with others, have been preserved until the present like relics of the ancient treasure.
>
> (Aristotle, *Metaphysics* 1074b1)[26]

Now let us see what we can say about the structure and function of these old "relics" based on our newly acquired knowledge about our ancestors' sophistication.[27]

---

[26] Aristotle (*Heavens* 270b9, *Meteorology* 339b27, *Politics* 1329b25); see also Werner Jaeger (1923b: ch. II, 6).

[27] [*Editors' note: This paragraph was followed by 22 manuscript pages of an explanation (section 7a) authored by Helmut Spinner, the then editor of the book. We decided not to include this text in the present edition, since it is not authored by Feyerabend himself. Spinner's research on this topic is included in Spinner (1977).*]

——— 2 ———

THE STRUCTURE AND FUNCTION OF MYTHS

[8] The first attempts to "rationally" explain mythological forms of thought to make them more accessible to an intellect already somewhat removed from them were conducted in the Western hemisphere by early interpreters of *Homer*'s work.[1] The debate about Homer, and thus the "ancient quarrel between [poetry] and philosophy" (Plato, *Republic* 607b5), starts with attacks by Xenophanes, who accuses Homer of defaming the gods:[2] "Homer and Hesiod have attributed to the gods all sorts of things which are matters of reproach and censure among men: theft, adultery and mutual deceit" (*Fragments* 21B11). Since, according to Xenophanes, "from the beginning all have learned according to Homer" (*Fragments* 21B10), Homer's defenders – especially the rhapsodists – were trying to find means and ways to uphold his former authority. One of these means was the attempt to either remove the offensive elements or explain them away by showing their "true" meanings as opposed to the first impression a reader may get from them. Thus, even in Xenophanes' lifetime Theagenes of Rhegium interpreted the gods as *forces of nature* and their quarrels (see *Iliad* XX, 55) as the interactions between those forces. Metrodorus of Lampsacus applied this procedure also to mythical heroes; he regarded Agamemnon as the ether, Achilles as

---

[1] For a concise overview of the numerous attempts to grasp the "core" of the myth, see volume IV of Wilhelm Wundt's *Völkerpsychologie* [Social Psychology] of 1920. See also Kerényi's 1967 anthology Eröffnung des Zugangs zum Mythos [Paving the Way to the Myth], the symposium *Myth* (Sebeok 1965, which discusses Müller's theory in detail), and the collection of essays *Myth and Mythmaking* (Murray 1960). A fairly recent proponent of the theory that at least the Greek myths are nothing but tall tales is Lobeck (1829).

[2] Regarding the debate about Homer see also Pfeiffer (1968: 25ff., 57, 59, 66, 73, 82ff., 94ff.) and the literature referenced there. See also Forsdyke (1964).

the sun, Hector as the moon, Helen as the Earth, Demeter as liver, Dionysus as spleen, and Apollo as bile. Athena, who restrains Achilles, became prudence, while Hephaestus was already identified with fire in Homer's own writings. Olympus became the image of an organism, and this image continued to be used later on to illustrate geographic proportions. It is supported in the etymology, traces of which can be found in Pherecydes' work (Cronus = *chronos*) and which flourished in the fifth century (see the examples in Plato's *Cratylus*, which are inspired by Democritus' Homer studies).

By contrast, Hecataeus of Miletus and the so-called logographers approach legends and myths with a very different intent. They aim to explore the obscure era of ancient history, analyzing legends and myths with the goal of discovering their *historic core*. The analysis is based on the assumption that each myth consists of two elements: an historic event and its interpretive exaggeration or distortion. Thus if we remove the exaggeration we obtain an historic report. An example: "Aigyptos did not come to Argos. According to Hesiod's poem he had 50 sons; *yet as I believe they did not amount to even twenty*" (Hecataeus, as quoted in Fritz 1967: 71). Or consider this example:

> The events of the story [about Geryon's cattle] took place at the Ambracian Gulf. The latter is located on the Western coast of the Greek mainland, that is, in Epirus, which in ancient times seemed like a faraway country at the very end of the world. When it was subsequently discovered that the edge of the Earth's surface was located quite a bit further west, the Geryon myth was also relocated to that more distant location, and it was eventually relocated even further west to an island in the Atlantic Ocean, although this final relocation may also have been motivated by a similarity of names. But people generally tend to exaggerate. This is why we need to retrace our steps to the realistic core of history. Moving the cattle of Geryon from the Ambracian Gulf to Sparta *is an astonishing effort in and of itself*.
>
> (Hecataeus, as quoted in Fritz 1967: 71)[3]

This *rationalist critique* of myths, which in Hecataeus' times gave history a new direction, later became an obsolete triviality in the work of Euhemerus. Aristotle emphasized the necessity of considering historical factors as well as of taking into account the different customs of the heroic era (*Fragments* 142–79; *Poetics* 1461a2). Aristarchus of Byzantium subsequently turned this into a basic principle for the interpretation of Homer's work (internal versus external

---

[3] A detailed discussion of Hecataeus' method and its philosophical background can be found in ch. III of Kurt von Fritz (1967). On the general topic, see also Grube (1965).

interpretation). With the attacks by Xenophanes (*Fragments* 21B11, 12; see also 14, 15, 32, 34) and Heraclitus (*Fragments* 22B40, 42) the once-uniform body of learning is split up into *philosophy*, which proceeds purely conceptually, aiming to eradicate the imagist way of thinking of earlier epochs, and *poetry* (verse, drama, etc.), which continues to employ the old tools even when presenting new ideas. *Both* claim to provide practical instruction, ethical guidance, and a *true report*.[4] *Both* criticize the "frivolousness" of Homer, the Homeridae, and the early myths. This odd agreement between the two rivals is due to their common dislike of fictional structures and to their difficulties with understanding *or even articulating* them, which appears to have been a special problem well into the classical period in Greece.[5]

2.1. Theories of Myth

[9] These early theories tacitly assume that Homer, the mythographers, and their more enlightened successors all share the same *experience* of the world but that, due to language, customs, deliberate obfuscations, exaggerations, or circumlocutions, experience was concealed and readers were misled. Once the narration's surface layer is accurately interpreted and its relation to nature or society (i.e., symbolic, allegorical, exaggerated, direct, etc.) correctly understood, it can be used to infer actual existing circumstances directly. Despite differences in intention and in manner of expression, inventors of myths share certain characteristics with our contemporaries: The former see the world, stars, plants, animals in the same way as the latter; they also perceive the same social patterns. For example, they see faces and facial expressions in the same way as we do. They have a comparable inner life (grief, anger, dreams are the same kinds of occurrences for them as they are for us), and they also react to external stimuli in the same way as a contemporary person (i.e., as a person in the times of Homer's first interpreters). This is the view underlying both the

---

[4] On poetry's truth claim, see Homer (*Iliad* II, 484–86); Hesiod (*Theogony* 22) – see also Gigon (1968: ch. 1), Stesichorus (*Fragments* 192f.), and Pindar (*Olympian* I, 35f.; X, 3). On Aeschylus and Pindar, see Schachermeyr (1966: chs. 8 and 14).

[5] See Forsdyke (1964: ch. 8) as well as comments in sections [17] and [18] [*chapter 4.1*]. On Theagenes, see DK 8A1 and Pfeiffer (1968: 26ff.). The Pherecydes fragments are thoroughly discussed in Rudolf Eisler (1910: ch. IV), which also contains an account of the method of isopsephy (1910: 340f.). See Wilhelm Nestle (1942: ch. 5) on the allegoric and rationalist approaches to the interpretation of myths as well as on the logographers' procedures. A non-allegoric interpretation of his ideas is provided by Cornford (1912: 18). See the relevant articles on any of these topics in Wissowa (1958).

rationalist and the allegoric theories. For it makes sense to speak of an "allegory," a "symbolic representation," or an "exaggeration" only if, alongside the allegoric, symbolic, or exaggerated description, there is or can in principle also be an "objective" description – and according to the aforementioned theories the latter is based on a collection of facts shared and perceived by us as well. Thus, allegoric, symbolic, and rationalist theories of myth are all just variants of the theory of *nature or society myths* according to which myths *start* with a core of correct and easily reproducible perceptions of the world and of humans, though these perceptions are subsequently *described* in different ways, either realistically (pure nature myths) or in an esoteric, exaggerated, ritually stipulated language (allegorically distorted nature myths). Even a desire to conceal a myth can contribute to its odd appearance.

[10] The theory of nature myths (where "nature" always includes society as well) is sometimes based on a naïve and at other times on a more refined *epistemology*. The *naïve* version, which is closely related to naïve realism, assumes that the elements of reality, the "facts," are unambiguously presented to human consciousness and that these facts can be unambiguously described with the help of concepts. Differences in concept formation or in the contents of myths are solely due to differences in the number and nature of the perceived facts. Furthermore, the facts are directly represented; that is to say, a myth involving crocodiles actually describes crocodiles. Concepts are unilaterally based on facts, whereas the facts themselves – or their representations in human consciousness – are in no way influenced by concepts, rites, institutional peculiarities, or social circumstances.

The *refined* version, by contrast, assumes that such influencing does occur and considers it an important component of our knowledge. We are no longer dealing with an independent variable called "reality," which determines the thoughts and perceptions of people living in it, but with a series of factors such as language, artistic means of expression, social structures, perception, religion, emotions, which all interact and whose interconnections at any given moment are insufficiently grasped by the existing concepts. As with the naïve variant, here too the mythological vocabulary is deemed an early "language of science": whereas Euclidean science used circles, squares, lines, points, and the like, the inventors of myths used *social* and *biological* concepts and managed to find concrete representations of them. In lieu of geometrical constructions, they use a story represented in pictorial images. Unlike the naïve version, however, this version assumes that, first, even the simplest facts are at least in part *shaped* rather than merely described by those concepts and, second, the facts

37

are not always directly represented. Stories involving crocodiles, for example, may represent the *structure* of astronomic or social events. In such a case the crocodile words are not invariable building blocks of allegories but *variables* enabling us to identify the fixed points of the structures presented.

The first assumption gives rise to the conjecture that the strange events described in the myths as well as the strange creatures with which they populate the world are *really perceived* in it. The world really appears to this early thinking as a "You," not as an "It," the sky as a "picture book" rather than a "computation book," and each phenomenon described *actually exists, is perceived*, and is accounted for accordingly.[6] There is nothing of symbolism, allegory, and conceptual playfulness. Instead, "appearances are directly apprehended just as they are represented by the myth. Thus, here the mythological idea is reality and not just a symbol" (Wundt 1920: 34). This would explain both mythical thought's connection to reality and its personal and, insofar as there is worship and adoration, religious component. Rites can be explained as components of the mode of representation used. Oddities in artistic style and ideology, which the naïve version sees as the results of a lack of knowledge or talent, of "primitivism" and "magical" thinking, can be adequately explained by this theory at least in principle, provided that we manage to *identify* the psychological mechanism at work here and to *examine* its function independently of the myths interpreted. The third chapter contains fragments of such an examination. The second assumption mentioned above prevents a naïve inference from the *events* of the myth and the words describing them to the *intended objects*. Just as a physical theory introduces models without being interested in all of the models' aspects, a myth may present social or zoological episodes in order to illustrate some general cosmological structures. In such a case the names of individuals and classes are not constants but variables, and the purpose is explaining a *structure rather than* reporting on *concrete events* exemplifying this structure.

The transition from the naïve theory of nature myths to the more refined version is paralleled in more recent developments in philosophy of science. Here, too, it was first assumed that scientific concepts and scientific theories are uniquely determined by the phenomena of nature and that any differences in their theoretical structure must be

---

[6] On "You" and "It" see Frankfort, Groenewegen-Frankfort, Wilson, et al. (1949: 12, 29). On the "picture book" versus "computation book" see Alfred Jeremias (1929: 16, 239). According to Jeremias (1929: 25f., ch. IV) and Jacobsen (1949: 137ff.) the *Sumerian worldview* is based on an exact correspondence of Heaven and Earth, city and universe. See also the wealth of material in Meissner (1925).

due to errors of thought or experiment. Subsequently people realized that concept formation, since it is guided by the imagination, can develop in free and independent ways, being reined in by a comparison with nature *only in retrospect*. And finally people realized that such comparison is not a unilateral matter but that the use of a particular conceptual apparatus, such as a symbolic system, an inventory of elementary stories of which all more complex events are composed, or an artistic style, *changes* the appearances of nature and thus the objective facts used in such control processes.[7] Thus the most recent theory of myths has a lot in common with the most recent theory of science. This justifies our comparing myth and science not only with respect to their external appearance but also with respect to their cognitive contents. Chapters 3 and 4 contain a hint of such a comparison page by page in close connection with the historical account.

The naïve theory is represented by all those inductivists who do not deride myths as mere preconceptions, as the results of a pathological process of thinking, but as intermediate steps in the inductive ascent to pure knowledge. An example is Edward Burnett Tylor's *Religion in Primitive Culture*. Tylor denies the existence of the allegorical and the symbolic variants of the theory of nature myths. According to him, mythological concepts are all object-based and the myths themselves are "a perfectly rational and intelligible product of early science" (Tylor 1873: 4). Gods, demons, ghosts, vampires, and their various destinies are not just "mere creations of groundless fancy" – that is, they are neither symbols nor analogies – but "causes conceived in spiritual form to account for specific facts" (Tylor 1873: 278) whose apperception is based "on the very evidence of their [i.e., the savages'] senses" (Tylor 1873: 62). "In nations where the theory of the firmament prevails, accounts of bodily journeys or spiritual ascents to heaven are in general meant not as figure, *but as fact*" (Tylor 1873: 157, my emphasis). Concepts and theories are introduced in order to solve "two groups of biological problems" by which "thinking men, as yet at a low level of culture, were deeply impressed":

> In the first place, what is it that makes the difference between a living body and a dead one; what is it that causes waking, sleep, trance, disease, death?[8] In the second place, what are those human shapes which appear to us in dreams and visions? Looking at these two groups

[7] On the transition from the naïve to the more refined theory of scientific knowledge see Lakato and Musgrave (1970) as well as Feyerabend (1975).

[8] Such as the process of becoming increasingly weaker in certain forms of illness, without there being a visible cause – according to Tylor the most significant evidence for vampires; see Tylor (1873: 277).

of phenomena, the ancient savage philosophers probably made their first step by the obvious inference that every man has two things belonging to him, namely, a life and a phantom.

(Tylor 1873: 12)

Theories and primitive myths have the very same *goal*: to collect and causally explain salient appearances. Even the *method* is the same: it consists in drawing conclusions from existing observations. The differences in the results are due to the limitations of the observation material available to the myths' original creators. Tylor's theory has interesting parallels to Aristotle's theory of the formation of universals (Aristotle, *Posterior* 99b25ff.; *Metaphysics* 980b25): "induction exhibiting the universal as implicit in the clearly known particular" (*Posterior* 71a7). According to this theory the process of how universals are "implanted" in the soul depends on particulars as well as on "more rudimentary universals" (100b2) that have already been implanted in the soul, *but on nothing else*: "rudimentary universals" are formed "when one particular has made a stand," and particulars "make a stand" because "the soul is so constituted as to be capable of this process." The argument suggests that the same particulars give rise to the same "rudimentary universals," and subsequently to the same "true universals," *in every human being*. Aristotle also claims that universals are not innate but implanted inductively (*Posterior* 100b4) and that the soul is "pure from all admixture" (*Soul* 429b19), that is, it "can have no nature of its own, other than that of having a certain capacity" (*Soul* 429b23). Thus, human experience is shaped solely by the particular objects that we encounter. There are no other factors. For example, neither race nor social context is relevant, except insofar as they provide us with certain particulars. Given identical physical conditions, a slave and a barbarian alien would have the same *experiences* as a noble Greek.

We find a brief and pointed description of the naïve theory's epistemological background in Hegel's work:

> If thought and phenomenon do not perfectly correspond to one another, we are free at least to choose which of the two shall be held the defaulter. [The theory at hand] where it touches on the world of Reason, throws the blame on the thoughts; saying that the thoughts are defective, as not being exactly fitted to the sensations and to a mode of mind wholly restricted within the range of sensation, in which as such there are no traces of the presence of these thoughts. But as to the actual content of the thought, no question is raised.

(Hegel 1830b: §47)

40

The question is, however, raised in the refined theory. Various authors who realized that "facts" are not simply "given," but are constituted on the basis of existing ideologies, have used rudiments of the refined version of the theory of nature myths. According to Whorf, for example, the grammar of a language contains a cosmology, and language itself "is not merely an instrument for *describing* events but rather is the *shaper* of these events" (Whorf 1956: 212*). Walter F. Otto writes, "We cannot commit a greater error than [. . .] by trying to explain the myth from its respective form of being with its peculiar experiences, necessities and needs. The opposite is the case [which neglects interdependency]: It is only through the myth that its being obtains the form from which the relevant impulses, necessities and needs derive" (Otto 1956: 271). *Structuralism* has had an important role in the formation of the refined theory. Clearly the refined theory enables us to compare science and myth and apply the same standards to each.

The myths describe astronomical, biological, and sociological events, and their astronomical content may include not only conspicuous periods of sun, moon, and planetary motion, but also knowledge of precession, though not in the mathematical form it takes in astronomy. For in a "picture" book, instead of a gradual progression there is a repeated dissolution of familiar configurations by way of catastrophes and transitions from one "era" to the next. We do indeed also find such transitions in the tradition of a Golden Age, its demise and possible resurrection. The harmony of Heaven and Earth is in part a fact of nature – at least this is how we would put it today – but it is even more a result of a conscious adjustment of social life to astronomical phenomena. Cities and buildings are designed according to astronomical plans. "Babylonian towers" constitute a link between the sphere of Heaven and the land. Ecbatana was surrounded by seven walls that were painted in the colors of the planets and so wide one could drive on them around the city in a chariot.[9] Astronomical constellations, the partitions on the surface of the moon, the arrangement of the various sectors of the firmament,

---

* [*Translator's note: this is the way Feyerabend's German rendering of the original Whorf passage should be literally translated back into English. The original passage on the page referenced by Feyerabend, however, is slightly different in meaning from the way Feyerabend presents it in German. It runs: "[language itself] is not merely a reproducing instrument for voicing ideas but rather is itself the shaper of ideas." Obviously, there is a difference in something's being the shaper of ideas and something's being the shaper of events.*]

[9] See Herodotus, *Histories* L98. For Nineveh see de Santillana and von Dechend (1969: 239f.).

and planet colors in turn reflect geographical and social relations. On Earth "each phenomenon encountered by Mesopotamians in their environmental world [. . .] has its own personality, its own willpower, its own unique, distinctive Self" (Jacobsen 1949: 146[†β]). This self *reveals itself* in the causal properties of an object – as for example in a flint stone (which gets chipped even under soft horn pressure) – though it is not identical with the sum of them. With respect to the common reed, the goddess Nisaba, whom the writer addresses and thanks after successfully completing a piece of writing, is not only the power behind the reed's characteristic features and growth. In addition, she – "who knows the meaning of numbers and is able to reveal their significance" (Jeremias 1929: 41f.) – also promotes the art of writing.[10] Enlil's wild and unpredictable nature does not reveal itself merely in storms but also in hostile army attacks (the destruction of Ur by the Elamites). Within the cosmic hierarchy, to which all powers are subjected to a greater or lesser degree, he represents the force that keeps this hierarchy as well as the democracy of the gods on the right track – yet he can also unpredictably break the very order that he maintains and create chaos; the force that stabilizes the order can do this only at the cost of occasional excesses, which are an intrinsic part of its nature. This is the way in which the phenomena are linked to one another and to explanatory principles that serve to unite seemingly unrelated factors under a common aspect. Is there a cosmic experience subject to Enlil's arbitrary nature, which is projected onto earthly matters via a god? Some authors have assumed this.

According to these authors there is some basic astronomical knowledge far older than the traditional history of astronomy and even the history of letters, knowledge that can be traced back to the late Paleolithic age and that was once available in an "international mythical language" (de Santillana and von Dechend 1969: 302). This basic knowledge either degenerated or was lost. Astrology is a late product of this degeneration process, but like other products of degeneration, it contains valuable remnants that can be used to reconstruct the form and content of the ancient knowledge. The book *Hamlet's Mill* by de Santillana and von Dechend contains such an attempt at reconstruction based on a wealth of material. According to it the original cosmology described *temporal rhythms* rather than

---

[†β] [*Translator's note: Here again, Feyerabend took some liberties in his translation of the original English passage into German, which is here translated back into English. The original English passage runs: "In the Mesopotamian universe [. . .] everything [. . .] had a will and a character of its own."*]

[10] See Meissner (1925: vol. II, ch. 13) for her relationships to other gods.

geometrical proportions. "Earth" meant the ideal plain established by the ecliptic; "animated world" meant the zodiac strip on both sides of the ecliptic to the tropics of Cancer and Capricorn. The "Earth" is "quadrangular," resting on "four pillars" or "four corners," namely the solstices and the equinoxes. Precession makes the constellations shift; they disappear beneath the equator, beneath the "dry Earth," and sink down to the netherworld, from which they reappear after a long period of time. Periodically a "new Heaven" and a "new Earth" appear. Netherworld, mythical rivers, and the journeys of the soul all have astronomical meanings. The literary material is supplemented by pieces of art or commodities like landmarks, whose significance for our understanding of the history of astronomy and astrology was pointed out as early as the turn of the century (Boll, Eisler). Thus, starting with literary products we can gradually approach the distant past by way of interpretations and hypotheses, and achieve an astonishing level of insight into the knowledge of our predecessors. And these insights are fully confirmed by the very different kinds of research tools used in archeology.[11] This should suffice to eliminate the symbolic (and allegoric) interpretation of *early* myths (whereas symbolically encoded myths were indeed created later on, in Hellenistic times).[12]

[11] The *naturalistic* theory of myths and other forms of knowledge is not the only more recent theory, and it is certainly not the most popular. Yet the other theories either presuppose this one or are special cases of it. Or else they provide an answer to the question that alone is of interest for us here, namely of how the myth helps us humans in our attempt to understand and control nature, and in what ways it obstructs this attempt. Our problem is "the relation between the objective world and the subjective world of people, as it is shaped in various cultures" (Boas according to Benedikt).

Tylor's *animistic* theory, for example, is but a special case of the theory of nature myths rather than an alternative to it. According to it, nature is populated with spirits – and the myth serves to establish this. The *symbolic (allegoric) theory* works in similar ways, since all it states, after all, is that the relation between the myth's narrative

---

[11] See section [4] [*chapter 1.2*] on Paleolithic observatories, and section [5] [*chapter 1.3*] on Alexander Marshack's hypotheses and investigations.

[12] It should be noted that the excessively one-sided reduction of complex myths to one or two simple astronomical facts, as it was practiced in the nineteenth century (see Dorson 1958), temporarily led to some kind of "astrophobia" among myth researchers. This can be seen in Phyllis Ackerman (1960), who received reliable support in the astronomical ignorance among many of her professional colleagues. But here, too, we are already on the way to recovery.

(or art or rite) and nature itself is not simple and direct but has to be encoded in a roundabout manner (and of course, in addition to that the symbolic theory also assumes that its imaginative clothing does not change the natural phenomena in the least, which brings the symbolic theory very close to the naïve version of nature myth). The *rationalist theory*, according to which myths are the result of conscious efforts rather than "organically grown," has very little to do with our question at hand. It explains how a myth *develops*, but it rarely makes a contribution to questions about the myth's *structure* or *function* (in philosophy of science such an approach would be assigned to the "context of discovery," meaning that it fails to provide information regarding both its meaning and its reference to reality). An extreme case of the rationalist theory is the idea that myths are *purely fictitious*, "plain tall tales," as Plato would occasionally put it (*Phaedrus* 377d) – and similarly Aristotle (*Poetics* 1460a19f.) and Pindar: "tales told and overlaid with elaboration of lies" (*Olympian* I, 28f) – or the even later idea that they serve for *entertainment* rather than education or description of nature (Eratosthenes, *Fragments* I A 17, 20, 21). Yet Plato recognized very well that even a tall tale is able to manipulate the audience's spirit to such an extent that they will soon perceive the world in light of the tale. And he used myths in his own writings with that very purpose. What is decisive is not the sequence of narrated *episodes* but the *structure* at the base of that sequence, with its tacit assumptions about mind, body, thinking, and feeling.[13] Following Plato, we can say that myths express the general by means of fictitious particular events.[14] This brings Plato into the vicinity of the refined variant of nature (and societal) myths.[15]

From what has been said here, *evolutionist theories* do not have much to contribute to our account. Consider, for example, Max Müller's degeneration theory. Like Xenophanes and Pindar before them, both Müller and his opponent Andrew Lang were astonished by the barbaric elements of Greek myth.[16] They saw in them a strange anomaly that called for an explanation. How is it possible

---

[13] See Aristotle's *Poetics*: "poetry is something more scientific and serious than history, because poetry tends to give general truths while history gives particular facts" (1451b5ff.).

[14] See also the comments in section [16] [*chapter 3.2*].

[15] See also Stewart (1905), who, however, uses confusing Kantian terminology to articulate this matter of fact.

[16] On the long-lasting controversy between Max Müller and Andrew Lang about the theory of nature myths see Dorson (1958). On the problem addressed above and Müller's attempt at an explanation etc. see p. 19.

that a civilized people such as the Greeks produced such unflattering stories of its gods? Müller solved the problem by tracing the names of the Greek gods back to their Sanskrit equivalents, construing the latter as linguistic responses to natural events by linguistically impoverished tribes. Tradition fragments these responses into parts of a whole, which in turn keep triggering new interpretations. *Our* question, however, is how the *result* of this alleged degeneration, this "malady of language," manipulates our consciousness and modifies our behavior toward the environment: How can a "sick" language assist healthy people in their attempt to understand and handle their environment?[17] The same question can be raised with respect to the psychoanalytic theory, which in turn merely explains how myths *originate* from individuals or groups without indicating how the originated product is used to process our environment.

This brief survey of the various theories of myth confirms my suspicion that the *theory of nature myths* is best suited for the purpose of this presentation. It also allows us to treat myth, philosophy, and science in a uniform manner. And finally, it is consistent with *structuralism*, the most progressive contemporary theory of myths.[18] Apparent inconsistencies between the two approaches can be removed by analysis in the following manner.

[17] Note that evolutionary theories are discarded not because of their historical character but because they convey only the vaguest and most trivial kind of information about the *results* of this development.

[18] My quotes here are from the following works by Lévi-Strauss: *Structural Anthropology* (1958), *The Raw and the Cooked* (1964), and *The Savage Mind* (1962a). Ch. XI of *Structural Anthropology* contains an excellent introduction to the method used by Lévi-Strauss in his research, as well as some results. *The Raw and the Cooked* provides an application of this method to 187 myths, preliminary results as well as numerous detailed methodological tips. Lévi-Strauss aims to show that all variants of the content that the myth allows for are indeed gone through. *All* available variants, not just one so-called "true" variant, have to be considered in the analysis (see Lévi-Strauss 1958: 217; 1964: 13). This means that both Sophocles and Freud should be included in our sources for the Oedipus myth. Their versions deserve the same credibility ranking as other, older, and "more authentic appearing" ones (see Lévi-Strauss 1958: 217). Historical explanations of alleged distortions and inconsistencies should not replace the structural procedure of analysis, even where the latter encounters some difficulties (see Lévi-Strauss 1964: 196): if myths express basic forms of thinking then these forms must be present even in the oddest contortions of the material, and random historical events cannot hide them. Thus Kirk's objection in *Myth* (1970: 73ff.) that information orally passed down has its own laws is not convincing. If Lévi-Strauss is right in his hypothesis then a myth orally handed down is subject not just to the laws of oral tradition but also to the structural constraints of the human mind, and these are present in all circumstances. Small, "non-structural" facts especially can become the basis for discovering new and comprehensive structures (see Lévi-Strauss 1958: 327; 1964: 35).

45

2.2. The Theory of Nature Myths and Structuralism

[12] In apparent contrast to the theory of nature (and societal) myths, Lévi-Strauss wrote, "The mistake of Mannhardt and the Naturalist School was to think that natural phenomena are *what* myths seek to explain, when they are rather *the medium through which* myths try to explain facts which are themselves not of a natural but logical order" (Lévi-Strauss 1962a: 95). And, more distinctly: "The truth of the myth does not lie in any special content. It consists in logical relations which are devoid of content or, more precisely, whose invariant properties exhaust their operative value, since comparable relations can be established among the elements of a large number of different contents" (Lévi-Strauss 1964: 240). Thus, the myth's *objective* is to clarify a certain order that Lévi-Strauss calls a logical order, while its *medium* consists in sequences of tales designed to draw our attention away from individual events and toward general, formal properties, thereby highlighting the outlines of the intended order by their own structure through strategically placed repetitions (Lévi-Strauss 1958: 213).

Now, we have already seen (section [11] – [*chapter 2.1*]) that the refined theory of nature myths not only *accounts for* such an indirect manner of representation but actually *requires* it as a more flexible model to promote our understanding of the relations between humans and their environment. Lévi-Strauss' claim that the elements of a myth, the "mythemes" (Lévi-Strauss 1958: 211), do not have an absolute, inalterable meaning, but rather one that is primarily *positional* (Lévi-Strauss 1958: 213),[19] is not inconsistent with the refined theory. It can even add to his thesis that, due to the absence of individual characteristics, animal species assume the role of variables[20] by pointing to analogous procedures in science (use of models, change of the concept of force from something concrete to something abstract, structural). The *difference* between Lévi-Strauss and the approach that we propose here, however, lies mainly in the fact that our approach located the indirectly represented order inside

---

[19] This means that the mythemes serve to mark certain places within the structure that the myth displays for the purpose of recognizing analogous points in related myths; see Lévi-Strauss (1964: 81).

[20] "[A]s the nature of an animal seems to be concentrated into a unique quality, we might say that its individuality is dissolved in a genus" (Bergson, as quoted by Lévi-Strauss 1962b: 93). See also Lévi-Strauss (1962a: 61) – the mistake of the classical theory of totemism is that it reifies a classification scheme.

the world itself, while for Lévi-Strauss it is in the observer's mind. According to Lévi-Strauss the facts of myth are "themselves not of a natural but of a logical order" (Lévi-Strauss 1962a: 95). And since this order occurs not only in practical instructions but also in mythology, which possesses relative autonomy (Lévi-Strauss 1964: 351) and does not have a tangible practical goal (Lévi-Strauss 1964: 9), it must consist in deep-seated laws, in constraining structures of the mind in the manner of Kantian philosophy (Lévi-Strauss 1964: 10), rather than structures of the world surrounding the mythologist.

As results of specifically *human* activity, to be sure, scientific theories, myths, and other results of human activity reflect human properties. Whatever a researcher does, his behavior is subject to certain psychophysical laws, and this will be reflected in the products that result from it. Yet it does not follow from this that those products may not also reflect *other* influences such as, for example, properties of the world surrounding us humans (just as a portrait reveals both the painter's skills and temperament and the appearance and character of the person depicted). Furthermore, we have to admit, especially within the framework of the refined theory, that the "facts" of the world are not just externally *reflected* by human ideology but are at least in part *formed* by it. A language that does not differentiate between main and subordinate clauses, that instead describes the world in terms of a juxtaposition of declarative sentences (as is the case in part with Homer's language as well as to a full extent with the language of Wittgenstein's *Tractatus*), will not be familiar with "substances," nor will it have any way to even articulate this shortcoming. To the speaker of such a language the world appears as a substance-less aggregate of more or less closely linked building blocks. Nonetheless, the very sentence "The world is an aggregate not based on any substance" would not be a logical sentence even if it could be expressed in the language under consideration. For a logical sentence presupposes that the constituting process has been fully successful and that there is no remainder. Of course, any possible remainders cannot be articulated within the language itself – but is their formulation within a *stable* language really the only way in which we can recognize the limits of an ideology? Languages tend to change, they degenerate or develop, and what at first cannot be uttered may become a fundamental principle, often just because the imperfection of the language's existing state is initially perceived in a manner that cannot be articulated linguistically.[21] Given the possibility of such

---

[21] See Achilles' speech in *Iliad* IX, 508ff, and the discussion in section [17] [*chapter 4.0*].

changes, even a complete coincidence of ideology and world would be but a mere serendipity rather than a logical truth.

It follows from this that the scheme of symmetrically arranged oppositions and mediations that Lévi-Strauss frequently discovers in his material can *and has to* be conceived as a cosmological scheme. Of course, considered in isolation this scheme is nothing but a system (as yet unused) for the arrangement of impressions. *But it is not considered in isolation.* Wherever we find this scheme it is part of a description of the real world, which is represented as a world containing such oppositions and mediations. If mythological thinking is always based on recognition of certain oppositions and leads to their gradual dissolution through mediation (Lévi-Strauss 1958: 217), then it thereby expresses *that there are transitions in the world*, where it remains irrelevant whether these transitions are implicit in the objective material without needing any additional ingredients or whether they first have to be immersed in that material by way of hidden mechanisms. It is *upon* completion of these mechanisms that the *objective world*, observed by humans engaged in tacit ordering activities, contains transitions, and it is up to the myth to establish this. Likewise, the claim that human mortality is related to the lack of human determinateness is a claim about the situation of humans within the world, even if this situation originally was created by humans themselves.

We certainly have to thank Lévi-Strauss for leading us beyond the naïve-literal account to a more abstract interpretation of myths that does not have recourse to wild allegories but is strictly bound to the subject matter. And yet increasing abstractness, increasing generality do not imply that we leave the realm of cosmology and enter that of logic, nor does the establishment of the presence of abstract structures within a myth show that there is no narrative of concrete events.[22] Thus, we see: the contrast between the refined version of the theory of nature myths and Lévi-Strauss' structuralism comes from certain *terminological* differences (such as that the general does not belong to cosmology but to logic), which in turn can lead to easily corrigible theoretical errors (such as that logic describes forms of the mind, not of the world). There is no fundamental opposition in their respective approaches to myths.

Lévi-Strauss' ideas are comparable to the new and more complex

---

[22] Lévi-Strauss does not do justice to this aspect of mythological thought. His students take his search for abstract structures even further; see Maranda (1972: 151ff.), especially the table on p. 156.

forms of empiricism that have developed in philosophy of science within the past ten to fifteen years. His critique of naïve naturalistic-utilitarian theories of myth, totemism, and other elements of preliterate societies has much in common with the critique of naïve empiricist theories of science. In both cases we have to ask how a certain social product refers to the nature that surrounds it; in both cases we recognize that the received theories (of science, of myth) contradict conspicuous – though not always familiar – facts. And the problem can be solved by taking perception to be more variable, our intellect to be more comprehensive, and methodological rules to be less strict than the orthodox views allow for. Indeed, it is now possible to compare myths and scientific theories and to apply the same standards to both. Hence it is surprising that Lévi-Strauss occasionally adopts a naïve naturalistic view in philosophy of science, especially in physics.[23] This is another reason why his own discovery strikes him as larger than it really is.

---

[23] On the state of philosophy of science see Lakatos (1970).

— 3 —

HOMER'S AGGREGATE UNIVERSE

[13] It is now time to supplement our general reflection with analysis of a concrete example. I have selected Homer's epics, together with the characteristics of the performing arts of the time, as an example. Rarely are these epic presentations treated at the same level as the myths. They are commonly regarded as already being too "advanced" and "rational." Nor do they possess the historical-narrative structure regarded by some authors[1] as a basic characteristic of myths and clearly noticeable in Hesiod. Homer's *Iliad* XIV presupposes such a structure, though it is not used in the composition of the epic. There are two advantages to this: the first is the existence of usable results showing us how language and arts constitute different worlds and different experiences of the world, thereby refuting naïve realism. The second is the fact that our reading Homer directly enables us to study the processes that change a complete configuration, and thereby to study the arguments that play into our *evaluation* of such changes. Of course, it could also turn out that there are no such arguments, or that there are but without any impact.

One world dissolves and is replaced by another. How does this come about? Does reason have any part in the process, or do we over-estimate its power if we conceive of the philosophers' era as an era of growing rationality? Anthropological field studies can give us insights into the status quo or into the dissolution of mythical worlds; still, they cannot show us the indicators of an *independent development* into other fully developed forms of thinking and perception. This is the most important reason why, rather than relying on the results of such field studies, I make use of studies in classic philology, classic

---

[1] Among them Lévi-Strauss (1958: 1994; 1964: 30).

archeology, and history of art, and why I focus on the Greek gods rather than African or Asian or South American demons.[2]

3.1. The Paratactic World of Archaic Art

[14] The *archaic style* as it is defined in the first chapter of Emanuel Loewy's book *The Rendering of Nature in Early Greek Art* (Loewy 1900b) has the following characteristics.[3] (1) The structure and motion of figures and their components are limited to just a few *typical schemata*. (2) Individual forms are *stylized*; they are regular and executed at a highly precise level of abstraction. (3) A form's representation is based on its *outline*; the latter is either a line or a filled-in boundary, a silhouette. Silhouettes may be of various postures; standing, marching, rowing, driving, fighting, dying, wailing. Their basic structure is always clear. (4) Only one *hue* is present; there are no grades or shades. (5) The figures commonly display their parts (and more comprehensive episodes their episode components) from their geometrically best side, even if this interferes with their composition and leads to a distortion of spatial relations. (An extreme example of this is the charioteer in figure 11, who stands not within but above the side of his vehicle, so that his feet are visible. Late Mycenaean depictions such as the one in figure 12 are closer to our view.) (6) Corresponding to (5), *overlaps* tend to be avoided, and objects standing behind one another are represented as standing side by side (see the position of the dead body between ceiling and bier in figure 13). (7) An action's *environmental setting* (mountains, clouds, trees, etc.) is either entirely missing or indicated only schematically. Actions consist of units that are *self-contained* and represent *typical scenes* (battles, funerals etc.).

Loewy explains these stylistic elements, which we find in only slightly varied form in children's drawings, in Egyptian "frontal

---

[2] [Editors' note: In the typescript section [13] was not at the beginning of the third chapter but at the end of the second.]

[3] Loewy uses the term "archaic" to denote a certain concept that is supposed to cover certain phenomena in Greek and Egyptian art, as well as in the art of preliterate tribes. For Greece his work applies especially to the late geometric style (after 900 BC) and to the archaic period in the narrow sense (*c*.700 BC to 500 BC), which deals with the human figure in more detail, involving it in lively episodes. The present text accommodates the term "archaic" both in Loewy's meaning and in the more narrow, technical sense. The context will always indicate the difference in meaning. On Greek art see Matz (1950), Beazley and Ashmole (1966), as well as Pfuhl (1923), which is still worth reading. See also the relevant chapters in Schachermeyr (1966). Schäfer (1963) covers Egyptian art like no other.

11. Charioteer with chariot: detail of figure 18

From: Webster, Thomas B. L. (1958): *From Mycenae to Homer*, London, conclusion, fig. 22.

art," and in early Greek art, in terms of psychological mechanisms. The artists imitate not the object but their memory image of the object, and it is the latter that has the properties indicated. The style changes as a result of planned observations that modify the memory images. It is unclear, though, why the artists copy the memory images and not their perceptions, since the latter are so much clearer and more enduring, after all. Furthermore, it turned out that the realism that was supposed to follow after the archaic period of art, according to Loewy, often actually *preceded* the more schematic

12. Late Mycenaean vase with chariot, from Encomi

From: Webster, Thomas B. L. (1958): *From Mycenae to Homer*, London, fig. 15.

13. People resting: fragment from an Attic vase, 9th to 8th century BC

From: Beazley, J. D. and Bernard Ashmole (1966): *Greek Sculpture and Painting: To the End of the Hellenistic Period*, Cambridge, fig. 2.

forms of representation. This goes for the Paleolithic period,[4] whose realistic phase was not known at the time of Loewy's work, as well as Egyptian art and Attic geometric art.[5] Webster writes about Attic art:

> Attic Geometric art should not be called primitive, although it has not the kind of photographic realism which literary scholars appear to demand in painting. It is a highly sophisticated art with its own conventions, which serve its own purposes. [. . .] a revolution separates it from late Mycenaean painting. In this revolution figures were reduced to their minimum silhouettes, and out of these minimum silhouettes the new art was built up.
>
> (Webster 1958: 205)

In Egypt this development can particularly be seen in the depiction of animals. The so-called *Berlin Lion* is wild and threatening, very different from the majestic animal of the 2nd and 3rd Dynasties. The latter is not as much a representation of an individual lion as it is of the very *concept* "lion". The falcon undergoes a similar development from the animated form on King Narmer's palette (figure 14) to the majestic-conceptual form on the stela of King Djet (figure 15). "There

---

[4] On the change from realism to formalism in the Paleolithic period see Paolo Graziosi (1960) as well as André Leroi-Gourhan (1967), both with excellent illustrations. There is a brief and detailed account in chs. X and XI of Miles Burkitt (1963).

[5] [*Editors' note: In the typescript the following remarks on the development of Egyptian and Attic arts, together with figures 14 and 15, are located right at the end of section [14], just before Feyerabend turns to the Homeric epics.*]

14. Palette of the Predynastic King Narmer, c.3000 BC

From: Schäfer, Heinrich (1963): *Von ägyptischer Kunst: Eine Grundlage*, Wiesbaden, tablet 5.1, reverse side.

15. Stela of King Djet, 1st Dynasty, c.2870 BC

From: Webster, Thomas B. L. (1958): *From Mycenae to Homer*, London, fig. 15.

16. Vase in geometric style

From: Schachermeyr, Fritz (1966): *Die frühe Klassik der Griechen*, Stuttgart, tablet 13.

17. Funeral vase in Attic style, *c.*750 BC.

From: Webster, Thomas B. L. (1958): *From Mycenae to Homer*, London, fig. 21.

was a transition to clarity everywhere, and the forms tightened up" (Schäfer 1963: 15).

In all of these cases the archaic style in Loewy's sense is the result of a *conscious effort* (which, of course, can be supported or obstructed by unconscious tendencies or physiological regularities). Thus, instead of asking for the psychological causes of a style we should rather try to discover its *elements*, understand these elements' *functions*, and look for related traces in literature, everyday life, and ideologies (religion, philosophy, cosmology) in order to reconstruct the *worldview* in which the style was at home and whose traces it may even have helped shape. With respect to the late geometric style in Greece, which was Loewy's basis, this kind of investigation yields surprising and interesting results.

This style, which combines figures and ornaments (figures 16, 17, 18), is strictly conventional. The figures are executed "with mathematical precision" according to Webster (1958: 292), who sees in this an anticipation of rationalism and philosophy: heads with distinct chins, thin necks, triangular bodies, thin arms and legs (figures 17, 19a, 19b), no individual features, no types (such as old man, fat man, or the like). Almost all of the components are displayed in profile; they are *sewn together* like the parts of a mannequin instead of integrated into an organic whole. This "additive" aspect of the late geometric and even the archaic style can be most clearly identified in the treatment of the *eyes*. The frontal eye, which is already present in Mycenaean vases (figure 20) but disappears in the early geometric silhouettes, reappears in the profile drawings and continues on into the sixth century in both serious pictures (figure 21) and humorous ones (figure 22). It is not an *organ* of the body that participates in its movements, guides them, "sees" for them, but an item in an enumeration, a *list*: "and in addition to their feet, arms, legs, torso, and head, humans also have eyes." What goes for physical parts also goes for parts of a situation or an event. The states of a body (vital, sick, dead) are expressed not "organically" in special arrangements of its parts but in bringing the standard body with its standard arrangement of arms, legs, etc. into typical *positions* without changing them at all. The dead person from the crater in figure 18 is furnished with two arms, two legs, a triangular torso, and a head in profile; indeed, this is simply a standing human being in the full force of his life, though turned around at a 90 degree angle. In addition to exhibiting all aspects of a living being he is also placed in a death *position*, while being accommodated in the space between ceiling and deathbed.

One extreme example is the depiction of a deer that is half-devoured

18. Attic funeral vase, *c.*750 BC

From: Webster, Thomas B. L. (1958): *From Mycenae to Homer*, London, fig. 22.

19a. Shipwreck: Attic jug and detail

From: Webster, Thomas B. L. (1958): *From Mycenae to Homer*, London, fig. 28a, b.

19b. Shipwreck: Attic jug and detail

From: Webster, Thomas B. L. (1958): *From Mycenae to Homer*, London, fig. 28a, b.

57

20. Mycenaean warrior vase

From: Webster, Thomas B. L. (1958): *From Mycenae to Homer*, London, fig. 7.

21. Board-game players on an amphora by Exekias

From: Schachermeyr, Fritz (1966): *Die frühe Klassik der Griechen*, Stuttgart, tablet 34.

22. Detail from a water jug, eastern Greece, *c.*540 BC

From: Beazley, J. D. and Bernard Ashmole (1966): *Greek Sculpture and Painting: To the End of the Hellenistic Period*, Cambridge, fig. 46.

by a lion.[6] The lion is wild, yet the deer is peaceful, and the lion devours the deer – this sequence of ideas is displayed in the image as a peaceful-looking deer that has half-disappeared inside the lion's throat. In other words, the act of devouring has been *simply added* to the wild lion and the peaceful deer. Thus we have here what is called a *paratactic aggregate*. Such an aggregate's elements are all of equal importance, their arrangement is that of a simple succession, components are not affected by the presence of other components, and there is no hierarchy. It almost seems like we need to *read* the image rather than *see* it: a wild lion, a peaceful deer, the lion devouring the deer.[7] For unanimated objects the aforementioned process of representation results in a *replacement of spatial relations by visible lists*: the chariots in the funeral procession of figure18 come with wheels *plus* floors *plus* feet and legs of the driver (compare figure 11), while in the late Mycenaean period everything is covered that cannot be seen in reality (figure 12). Even statues are composed of their parts, which look like independent elements, and the transition from the side view to the frontal view, which are both relatively complete (the profile is a complete profile and corresponds to the profile view in painting), is abrupt; that is, both views are simply strung together as parts of the object (figure 23).

No doubt the additive interpretation, especially in the case of two-dimensional representations, has to be *learned*; it cannot be read off the image without any preparation. Examples of Egyptian drawings and paintings, which often can be decoded only in the presence of the object depicted, convey to us an idea of the nature and extent of the required information. Thus it turns out that the chair "24a" in figure 24 represents object "24b," and not object "24c." Figure 24a is to be read: "Chair with back and four legs, legs are connected by

---

6 [*Editors' note: Feyerabend does not provide us with an image here but only with a general reference to Hampe (1952). Hampe's book does contain depictions of lions and deer (tablets 12, 13, 14), but these would require a very generous eye to allow for Feyerabend's interpretation.*]

7 That the images should be more *read* than *seen* also goes for Egyptian art: "Indeed we can grasp the topic of drawings of bodies based on frontal images [i.e., archaic in Loewy's sense] best by first reading off the various parts of their content in enumerative statements [. . .] O. Wulff has occasionally used the term 'seeing concept' for such drawings" (Schäfer 1963: 118; see also Wulff 1927: 43–50). Webster, too, speaks of the "narrative" and "explanatory" element of Mycenaean and geometric art (Webster 1958: 202). Gronewegen-Frankfort offers similar observations: the scenes from everyday life covering the walls of Egyptian graves "[should] be 'read'. Harvesting entails ploughing, sowing, and reaping; care of cattle entails fording of streams and milking; boating entails fights between sailors; etc. The sequence of the scenes is purely conceptual" (Gronewegen-Frankfort 1951: 33).

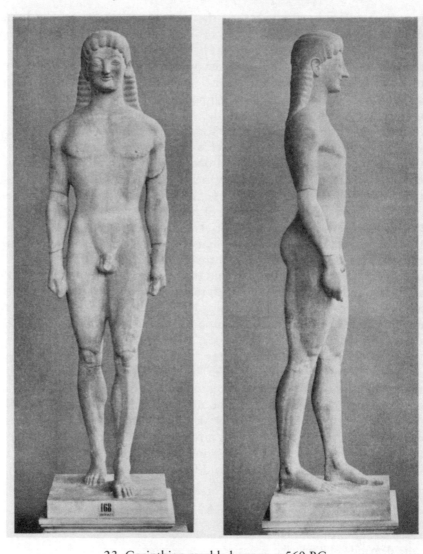

23. Corinthian marble kouros, *c.560 BC*

From: Beazley, J. D. and Bernard Ashmole (1966): *Greek Sculpture and Painting: To the End of the Hellenistic Period*, Cambridge, fig. 29.

supports," where our knowledge of the supports' position is tacitly presupposed.

The interpretation of groups is very complicated, and there are cases that still have not been resolved. For example, figure 25 suggests the arrangement of figure 26a, yet the correct arrangement is that of figure 26b. The ability to "read" a certain style also requires us to know what is relevant and what is not. Not every property on an archaic list has representational value. The Greeks overlooked this feature when they attempted to find out the reasons for the "dignified posture" of the Egyptian statues. That very question "might have struck an Egyptian artist as it would strike us if someone inquired the age or mood of the king on the chessboard" (Gombrich 1960: 134). With this we conclude our account and illustration of the archaic style according to Loewy.

A style can be described and analyzed in different ways. Up to now we have focused on the *formal aspects* of the archaic style: the style provides us with *visible lists* of the represented situation, the components of which are arranged roughly in the way in which they occur in nature, except when the natural arrangement hides important elements. The items on the lists are all equally important; they are "sewn together" rather than integrated into an organic whole. One item's content does not depend on another's – even the presence of a lion and the act of devouring a deer do not affect the peaceful appearance of the latter. Archaic images are paratactic aggregates rather than hypotactic systems. The elements of the images can be physical components of objects or bodies, heads, arms, wheels, or they can be states such as the state of being dead, or actions such as the act of devouring something. Another form of description is the *ontological* one: we describe the characteristics of a *world* that is built up just like the world of an archaic image, as well as the impressions that an observer of such a world receives. This form of description is especially popular with art critics. We can find it, for example, in Hanfmann:

> No matter how animated and agile archaic heroes may be, they do not appear to move by their own will. Their gestures are *explanatory formulae* imposed upon the actors from without in order to *explain* what sort of action is going on. Another crucial obstacle to the convincing portrayal of inner life was the curiously detached character of the archaic eye. It shows that a person is alive but it cannot adjust itself to the demands of a specific situation. Even where an archaic artist succeeds in denoting a humorous [see figure 22] or tragic [see figure 21] mood, these factors of externalized gesture and detached glance recall the exaggerated animation of a puppet play.
>
> (Hanfmann 1957: 74)

An ontological description, or a description of the impressions of an observer who suddenly sees himself or herself transferred to a world that precisely corresponds to the image, is commonly nothing

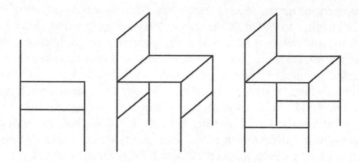

24. Chair perspectives a, b, c
Drawing by Feyerabend

25. The deceased King Niuserre among gods, with the king's guard at the bottom

From: Schäfer, Heinrich (1963): *Von ägyptischer Kunst: Eine Grundlage*, Wiesbaden, tablet 12.1.

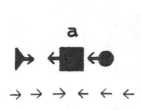

26a. Incorrect key to figure 25

From: Schäfer, Heinrich (1963): *Von ägyptischer Kunst: Eine Grundlage,* Wiesbaden, p. 223, drawing for tablet 12.1.

26b. Correct key to figure 25

From: Schäfer, Heinrich (1963): *Von ägyptischer Kunst: Eine Grundlage,* Wiesbaden, p. 223, drawing for tablet 12.1.

but a wordy and subjective-sentimental reintroduction of the most important aspects of a purely formal analysis. And yet we should not rule out the possibility *that a certain style represents the world in precisely the way in which it was seen and experienced by the artist and his or her contemporaries, and that basic assumptions (conscious or subconscious) are expressed in each formal characteristic, assumptions that reflect the cosmology of the time.* Thus, in the case of the 'archaic' style we should not rule out the possibility that the world of the archaic humans was indeed an aggregate of parts rather than an organic unit, and that archaic humans saw their fellow humans as loosely connected mannequins driven solely by external factors. Such a "realistic" interpretation of the archaic style would match Whorff's hypothesis that means of representation do not just serve to *describe* events – so that the latter could also have other characteristics not captured in the description – but actually *constitute* these events.

On the other hand, the realistic interpretation of a style is not a given. There *are* technical failures. Lack of competence, special purposes such as caricature, and deviations from a "faithful rendering of nature" often occur alongside more realistic representations and with a precise knowledge of the object: the studio of Thutmose in Tell el-Amarna (the former Akhetaton) accommodates masks of live models displaying all the details of head formation (elevations and depressions in the bony envelope) as well as heads corresponding to these masks. In some of the heads the details are preserved, but in others they are eliminated and replaced by simpler forms. An extreme example is the completely smooth head of a public servant that bulges way out in the back (figure 27). This proves "that at least some of the

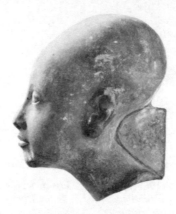

27. Head of a public servant from Amarna

From: Schäfer, Heinrich (1963): *Von ägyptischer Kunst: Eine Grundlage*, Wiesbaden, tablet 49.1.

artists intentionally worked independent of nature" (Schäfer 1963: 63). Two changes occurred in the manner of representation during the reign of Amenophis IV (1364–1347 BC). The first, which introduced a more realistic style, occurred only four years after his ascending the throne. Thus, the technical skills for realism already existed, but they were deliberately not applied or developed further.

Drawing conclusions from a style or language to a particular cosmology and perception requires meticulous arguments; it is not a matter of course and can easily mislead us. It requires proof that the characteristics of a certain style occur not only in images and sculptures but also in cultural areas entirely separate from fine art. For example, proof of the existence of philosophical doctrines that build up the world in *thought* just as well as art does in *images* would release the style in question from the narrower domain of art and show it to be an expression of a form of life. Such proof is indeed available for the archaic style in Loewy's sense. The first step in this proof is an analysis of the literature at the time, i.e. of Homer's epics.[8]

---

[8] The formal structure of Homer's epics has been reinvestigated in our century by Milman Parry, who also looked into the connection between the form and the requirements of oral literature. Parry's method and findings are discussed in ch. VI of Page (1959) as well as in articles in Kirk (1964) by Dodds, Grey, Lord, Kirk. See also ch. I of Kirk (1965). All of these works include detailed additional references as well. The relation to Near Eastern formulaic language and archeological find is discussed in Webster (1958); see also Whitman (1958). Gilbert Murray (1934) is still worth reading and full of ideas and suggestions. Finley (1970) contains a summary of the historical background.

3.2. Worldview and Knowledge in Homer's Epics

[15] Analysis of the structure of Homer's epics has yielded interesting results in this century. The Nine tenths of the epics' content consists of *formulas*, that is, of already existing phrases that are inserted at suitable places and frequently repeated. The formulas are of various lengths. They can extend over several lines: *Iliad* IX, 17–28 is copied from start to finish from II, 110–41; IX, 14–15 corresponds to XVI, 3–4, and I, 23–32 corresponds to I, 60–90. The principle here seems to be that a phrase already formulated in a satisfactory way gets *literally* repeated in a similar situation.[9] Repetition of details such as introductions to speeches, forms of addressing an individual, and description of battle deaths and other routine events is more frequent. Most interesting is the repetition of *partial* lines, and especially, repetition of noun–adjective combinations. One fifth of all lines get fully repeated in one or another location, and there are approximately 25,000 repetitions in the 28,000 lines. Repetitions, once decoded, can be traced back from Linear B via Mycenae to oriental prototypes, especially to the formalities of aristocratic-bureaucratic societies and to court literature influenced by them.

> Titles of gods, kings, and men must be given correctly, and in a courtly world the principle of correct expression may be extended further. Royal correspondence is very formal, and this formality is extended beyond the messenger scenes of poetry to the formulae used for introducing speeches. Similarly operations are reported in the terms of the operation order, whether the operation order itself is given or not, and this technique is extended to other description, which have no such operation order behind them. These compulsions all derive ultimately from the court of the King, and it is reasonable to suppose that the court in turn enjoyed such formality in poetry.
>
> (Webster 1958: 75f.)

---

[9] Examples of repetitions in letters: a letter of Shamshi-Adad, king of Assyria, to his son Yasmah-Addu, king of Mari, starts as follows: "Speak to Yasmah-Addu: thus (says) Shamshi-Adad, your father, regarding the Telamonian messenger of whom *you wrote to me* thus: 'he entered a merchant's home [. . .] *and someone beat him*. This is why I *still have not sent him to you*.' This is *what you wrote to me*. Now assume that *someone beat him*. Can he not ride a horse? You *still have not sent him to me* . . . Why?" In the epics the style of exact repetition reoccurs as exact repetition of question and answer. (This style still exists today, e.g., in Stalin's later didactic writings.) Directives for military operations and for religious sacrifice provide a literal repetition of long phrases. Example: Pylos Tablet Ta 326: "Pylos: Perform a certain action at Pa-ke-ya-ne and bring the gifts and bring those that we carry. To the Mistress: a golden bowl, a woman" – and this is repeated four times in different locations and for different gods.

Conditions at the (Sumerian, Babylonian, Hurrian, Hittite, Phoenician, and Mycenaean) courts also explain why certain standard elements of *content* (typical scenes, the king in war and peace, furniture, beautiful objects, jewelry, battle, and armor) constantly get repeated. They are at the center of attention; people speak of them often and are pleasantly affected by them. This explains the invariable features of both form and content. The rest varies according to local circumstances; episode details names, and geographical descriptions change while the general content-related framework, as well as the formal features, remains the same. Even the much more moderate variations characteristic of Mycenaean heroic ballads, which have to meet expectations at various courts and of various guests at the same court, have the same invariable background. This is how the *combination of invariable and variable elements* came about that would be so important for subsequent development.

This combination is then utilized by those poets who, exiled to foreign coasts in a preliterate age due to the chaos of the "Dark Ages," would sing ballads about their Mycenaean past. Relying solely on their *memory* they created a language and a method of expression that was best suited to oral composition and its preservation in memory. This language adopted the court formulas, documents, and catalogs that poets rarely leave behind and that they use under compulsion, not by free choice. Along with the *social* form of compulsion there was the additional *compulsion of rhythm*, which was even more important for recitation and required that a phrase fit the lyrical meter, as well as memory-relevant considerations of economy reducing the number of formulas that needed to be learned to a minimum. Where meaning and rhythm clash the latter often prevails over the former.

Note how Zeus changes from counsellor to storm-mountain-god to paternal god, *not* in connection to what he is doing, but at the dictates of metre. He is not *nephelegerata Zeus* when he is gathering clouds, but when he is filling the metrical unit vv-vv--. Just so, Achilleus, whose name is the exact metrical equivalent of Odysseus (v--), shares with him the rather general epithet *dios*. And Menelaos and Diomedes get the odd epithet "of the great war cry" not because they are noisier than anybody else, but because their metrically identical names can usefully terminate a line beginning with ---vv-v with *boen agathos* (Menelaos), *boen agathos* (Diomedes). Now such an epithet is (metre aside) neither specially appropriate nor yet inappropriate. Elsewhere, though, the formula trails its adjective into surprising contexts. So Aigisthos, of all people, is "blameless". Aphrodite is "laughing" when she voices a complaint (*Illiad*, V, 375) in tears. If Helen will be the "beloved wife" of the rival who wins her (*Illiad*, III, 138) this implies no archness or slyness on the part of the speaker. "Beloved" goes [metrically] with "wife".

(Lattimore 1951: 39f.)

Thus we see that "the poets have no choice, do not even think in terms of choice; for a given part of the line, whatever declension-case was needed and whatever the subject-matter might be, the formular vocabulary supplied at once a combination of words ready-made" (Page 1959: 242). Along with the compulsion of rhythm goes the compulsion of economy.

"All the chief characters of the Iliad and Odyssey, if their names can be fitted into the last half of the verse along with an epithet, have a noun-epithet formula in the nominative, beginning with a simple consonant, which fills the verse between the trochaic caesura of the third foot and the verse-end: for instance, [*polutlas dios Odysseus* ...] In a list of 37 characters who have formulas of this type, which includes all those having any importance in the poems, there are only three names which have a second formula which could replace the first" [Parry 1930: 86f.]. Thus equipped the Homeric poet "has no interest in originality of expression [. . .]. He uses or adapts inherited formulas."

(Page 1959: 224)

Now, did this result of an *external* requirement like the requirement of easy reproducibility in one's memory have an impact on perception, on general ideas about world and man, or on fine art? And did such an impact in turn result in certain features of composition related to those of "archaic" art in Loewy's sense? These are some of the questions that a *sophisticated naturalist* must ask in this context. Looking at the following additional features of epics will make it easier for us to find answers to them.

Homeric poets use their formulas to represent typical scenes in which objects are occasionally described "by adding the parts on in a string of words in apposition" (Webster 1958: 100).[10] Just as with the geometric style, the elements are "sewn together" rather than combined into an organic whole. This goes both for *physical* components and for the elements of complex *events*: ideas that we today assign to different

---

[10] Typical scenes in Homer and their Near Eastern prototypes are discussed in Webster (1958) and Whitman (1958). The latter work also contains information about the way in which more comprehensive scenes were composed from less comprehensive ones, comments on the macrostructure of epics, and a comparison of these with pottery of the time. See also Arend (1933). Kurz (1966) offers us a detailed discussion of minimum scenes of *motion*.

logical levels were narrated in a sequence in a number of grammatically equivalent sentences.[11] The following is an example: Meleager "abode beside his wedded wife, the fair Cleopatra, daughter of Marpessa of the fair ankles, child of Evenus, and of Idas that was mightiest of men that were then upon the face of earth; who also took his bow to face the king Phoebus Apollo for the sake of the fair-ankled maid. Her of old in their halls had her father and honoured mother called Halcyone by name" (*Iliad* IX, 556f.), and so forth through three other topics until the end. This *paratactic* feature of Homeric epics, which is perfectly understandable given the absence of a developed system of main and subordinate clauses in the early Greek language, also explains why Aphrodite is described as "laughing" when she voices a complaint in tears (*Iliad* V, 375), and why Achilles is still called "swift of foot" when he sits in conversation with Priam (*Iliad* XXIV, 559). Just as a corpse in late geometric art is still a living body that has been placed in the death *position* – the fact of death is *simply added* to the facts concerning the buildup of the body – and just as a lion's devouring of a deer is represented by positioning the deer (as a *peaceful*-looking animal) in a suitable relation to the lion's mouth – it is inserted in the image at the right place – so is teary Aphrodite nothing but Aphrodite – and that means, the *laughing* goddess – *inserted* into the situation of complaining.[12] She occurs in the right place in this situation, but she participates only externally; it does not affect her own nature.

This *additive* treatment of objects and events becomes especially clear in the comparisons that cut the human body into strictly separate units: Hippolochus' torso rolls like a round stone amid the battle throng after Agamemnon shears off his arms and head with a sword (*Iliad* XI, 146); Hector was sent whirling like a *top* (*Iliad* XIV, 412); the head of a fatally wounded warrior tilts like a poppy (*Iliad* VIII, 302). Even a continuous movement is broken up into individual events much like in a filmstrip. In *Iliad* XXII, 401, Achilles drags Hector, "the dust rose up, and on either side his dark

---

[11] On the characteristics of the early Greek language see Kühner, Blass, and Gerth (1966). On the development of the Greek language from paratactic side-by-side to sentence periods with superordinate and subordinate clauses see Webster (1957), as well as Kurt von Fritz' review in *Grundprobleme der Geschichte der Antiken Wissenschaft* (1971: 512ff.).

[12] Examples of concrete titles are: "Gilgamesh, Lord of Kullab" in Sumerian; "Kumarbi, father of the gods" in the Hittite song of Ullikumi; "Baal, rider of the clouds" in Ugaritic; "Zeus, father of gods and men" in Homer. Despite the absence of religious compulsion we encounter formal titles already in the omniscient official Mycenaean documents: "The Count Alectryon, the Eteoclean"; "Hector, slave of the gods"; "Petica, fuller of the king"; etc. On the Mycenaean documents see Ventris and Chadwick (1956). On the entire topic see Webster (1958).

hair flowed outspread, and all in the dust lay the head" – the *process* of dragging thus contains the *state of lying down* as an independent component that together with other components constitutes the motion: for the poet time consists of moments (Kurz 1966).[13] We see how the poet repeats the formal characteristics of late geometric and early archaic art step by step. Typical scenes, presented from their best side, follow one another as independent components of the total situation without any hierarchical structure and without an "underlying substance" that would enable their fusion into a "higher entity."[14]

[16] The absence of an "underlying substance" in the *formula* apparatus corresponds to the absence of *concepts*, which would conceptually integrate the various body parts into a living, organized unit, and its ideas, actions, moods, and attitudes into a "soul." There is no expression for the human body as a whole. *Soma* is the dead body, the corpse, while *demas* is accusative of specification and means roughly "with respect to form," "with respect to structure." Reference to limbs,[15] that is, to an aggregate, replaces our talk about

[13] This is, by the way, the theory that Aristotle (*Physics* 251b31) ascribed to Zeno and introduced in his argument of the flying arrow; this theory is a remnant from an older period of speculation that philosophers have tried in vain to integrate into their own worldview. The problem of the continuum, conceived of as a collection of discrete units, is still awaiting a satisfactory solution.

[14] The paratactic features of Homer's epics were one of the inspirations for Jakob van Hoddis' poem *End of the World* ("Weltende," 1911):

> World's End
> Whisked from the Bourgeois' pointy head hat flies,
> Throughout the heavens, reverberating screams,
> Down tumble roofers, shattered 'cross roof beams
> And on the coast – one reads – floodwaters rise.
>
> The storm is here, rough seas come merrily skipping
> Upon the land, thick dams to rudely crush.
> Most people suffer colds, their noses dripping
> While railroad trains from bridges headlong rush.
> [*Translator's note: Translation by Richard John Ascaráte.*]

"These two verses," wrote Johannes R. Becher, "these eight lines seemed to have turned us into different people, to have lifted us up from a world of dull bourgeoisie, which we despised yet of which we did not know how to leave it behind. These eight lines have captivated us [. . .] The experience of a new world seemed to hold us in thrall, an experience of simultaneousness of all that happens. Some learned literary critics soon found a name for this, namely simultanism. Yet Jakob van Hoddis taught us [. . .] that this experience of simultaneity was already inherent in Homer's work. For in Homer comparisons are not used in order to illustrate a point but in order to create in us the experience of simultaneousness, of unmeasured world vastness. When Homer described a battle and compared the clanging of weapons to a lumberjack's strikes, then he used this comparison only in order to show us that during the ongoing battle there also was simultaneously stillness in the woods, interrupted only by the lumberjacks' blows"

(Becher 1965: 52ff.).

[15] *Guia* = the limbs as moved by the joints; *melea* = the limbs with regard to their muscle power; *guia tromeontai* = his whole body trembled; *plēsthen d'ara hoi mele entos alkēs* = his body was filled with strength. See Snell (1946b: 5).

the one body and its modifications, while in other cases surfaces – such as *chros*: the body's surface – must do the job. Homer does not even have words for arms and legs. He speaks of hands, forearms, upper arms, feet, thighs, and calves. It seems that neither Homeric Greek thought nor its artistic representation acknowledged a body in the modern meaning of the word. At the most, a mannequin was conjured up that then got involved in a multitude of astonishing actions.

Nor did this mannequin have a *soul* in our sense. Just as the body is composed of limbs, parts of limbs, surfaces, and motions, so is the "soul" composed of "soul" events that are combined into interesting aggregates and sequences. Some of these events are not even of a "private" nature but the result of external interference by gods or inferior daemons. Basic processes and events are described according to quantity, not quality or intensity, and they are linked to one another like in a chain. "Never does Homer, in his descriptions of ideas or emotions, go beyond a purely spatial or quantitative definition; never does he attempt to sound their special, non-physical nature" (Snell 1946b: 18). Actions and even events of the "inner life" are not triggered by an "autonomy of the subject" that can be grasped neither by language nor by images. Rather, they are triggered by other actions including actions performed by the gods. "Persons do not yet possess an impenetrable exterior skin, and the divine is not yet something alien. Powers freely flow into the human being" (Fränkel 1960: 168). Accordingly, time is conceived of as an "approaching of happenings. We humans always look into the oncoming stream of time, which brings with it future happenings from a far distance" (Fränkel 1960: 14). These happenings interweave to form a tight network with no gaps or blanks. And this is precisely the way in which we *experience* our inner life. Dreams, as well as unusual psychological episodes such as sudden remembrance, sudden forgetting, rapidly growing rage, sudden increase of energy during a battle, and extraordinary changes, are not only *explained* in terms of divine interference but also *experienced* as intruders in the (usually unnoticed) routine course of events. Agamemnon's dream "went his way [and] came to the swift ships of the Achaeans" – the dream went his way, not just a character in it – "went his way to Agamemnon [. . .] sleeping in his hut. [And] took his stand above his head [. . .]" (*Iliad* II, 16–20). People do not *have* dreams (a dream is not a "subjective" event); rather they see them (dreams are "objective" events), and they can also see how dreams approach them and move further away again. (With some effort we can still reproduce these phenomena even today.) Thus, this experience of dreams is very different from our present one.

On one hand, the sudden increase of energy in the middle of a battle, which can occur either spontaneously or as a result of a prayer, is an objective process capable of occurring also in animals, and on the other hand, human beings experience it as an influx of energy coming from outside. The sentence "But as for valour, it is Zeus that increaseth it for men or minisheth it" (*Iliad* XX, 242) thus is not merely an objective description; rather, it also describes the experience of an external power acting on the person and filling him "with valorous strength" (*Iliad* XIII, 60). Today such occurrences are either quickly forgotten or brushed aside as random events. "[B]ut for Homer, as for early thought in general, there is no such thing as accident" (Dodds 1951: 6).[16] Everything is, in principle, explained. This belief articulates our inner life, it makes it clearer, it gives it an objective quality, and this objective quality again confirms the gods' intervention, which is used to explain startling mental events. It turns the gods into visible, tangible, sensible factors in the world: "The gods are there. To side with the Greeks in recognizing and acknowledging this as an accepted fact is the first requirement for an understanding of their beliefs and culture. Our knowledge that they are there rests on a perception, be it internal or external, and be the respective gods perceived directly or only via their recognizable effects" (Wilamovitz-Moellendorff 1931: 17).

To sum up: Homeric humans *were* far less compact than today's sensually nervous subject, and they also *noticed* it. They weren't compact in a *physical* sense; their bodies were composed of a multitude of limbs, surfaces, transitions, which – as we saw – are often taken out of their contexts when compared to, and described as, rollers, spheres, cones, or tops. They weren't compact in a *mental* sense, either, since their experiences are not related to an inner center of power, a "soul"; instead, they are simply added to the bundle of limbs, usually from outside: "Homeric man has no unified concept of what we call 'soul' or 'personality'" (Dodds 1951: 15). When Homeric humans cause events to occur, the events are not actually shaped by them; rather, they are complex arrangements of components with a blank space for which the individual in question is a precise fit. This can also be seen from the fact that "verbs for doing and making in Homer imply far less activity than the corresponding words that we use today. *Prattein* actually means 'to cover a distance,' but this is less about the effort that it takes an individual to accomplish this than

---

[16] See also the introduction to Walter F. Otto (1947b).

71

about whether we 'are well' while doing it. The Attic phrase *eu prattō*, 'I am fine,' confirms this" (Snell 1962: 47).

Success is not a result that flows from an effort and is formed by it; rather, *success is the lucky coincidence of a manifold of circumstances fitting together with the situation of the mannequin embedded in them*. These very concrete features of the Homeric world leave their traces even in the more abstract parts of their accompanying ideology. For example, Homeric (or "archaic" in Loewy's sense) humans are also ideologically "substance-less." They tolerate eclecticism in religion, accept foreign gods and myths without any hesitation, and let different versions of the same myth continue to exist side by side without attempting to remove any contradictions. There is no priest-hood and no dogma; a characteristic of this kind of everyday world is tolerance toward a diverse range of opinions about the world and its objects, absence of all theoretical exclusiveness, and absence of apodictic assertions about the world and the gods. (This tolerance can still be seen later in the Ionian culture, where ideas are presented as pleasant personal speculations without any purpose of triggering a conflict with traditional ideas. Only Xenophanes practices criticizing, ridiculing, and despising, thereby preparing the intolerant atmosphere of later philosophy.) There is no religious "morality" in our sense, nor are the gods abstract embodiments of eternal principles. They become this only at a later time, in the archaic era, as Eric R. Dodds has reported about Zeus as an example of the general development of the god concept: "in becoming the embodiment of cosmic justice Zeus lost his humanity. Hence Olympianism in its moralized form tended to become a religion of fear, a tendency which is reflected in the religious vocabulary. There is no word for 'god-fearing' in the *Iliad*" (Dodds 1951: 35). Thus "moral progress" leads to a loss of humanity.

Concepts of *knowledge* also reflected this lack of uniformity and compactness.[17] There was no comprehensive concept of knowledge, just as there was no comprehensive essence of things. Different words were used to express what we today regard as different *forms* of knowledge or as different ways to *obtain* knowledge. *Sophia* means experience in a particular craft or trade (e.g., as a singer, carpenter, general, physician, charioteer, freestyle wrestler), which includes those arts in which the artist is regarded not as an unique *creator*

---

[17] On the concept of knowledge see Snell (1924) and ch. IV of Snell (1946b). The term *sophia* occurs only once in Homer's work, namely in *Iliad* XV, 412, in reference to a "cunning workman" (i.e., a skilled carpenter).

but rather as a master who knows how to place the right thing in the right location (see above on the concept of *practice*). *Eidenai*, literally "having seen," refers to knowledge "to all appearances," while *suniēmi* implies (especially in the *Iliad*) the notion of following and obeying – not in the sense in which we *face* a thing, explore its features, and then act in accordance with these features, but rather in the much more direct sense of *fitting* into the (not always sensuously) perceivable environment.

> As long as our "understanding" of the world occurs as an original active development of imagination we are missing the experience of facing the world from a theoretical-observational (or here better "devoting-listening") point of view, and hence primitive humans cannot really recognize this attitude as a genuine "understanding." As we saw in the *Iliad*, not even the assumption of meaningful speech was regarded as "understanding" but only as a mere "following." For in the initial stage of the process there is not even a distinction between grasping and the thing being grasped.
>
> (Snell 1924: 50)

Just as the act of devouring in late geometric art is represented not as a transformation of what is being devoured but rather as a correct juxtaposition of devourer and devoured, so too is knowledge obtained by a correct juxtaposition of knower and known. The muses in *Iliad* II, 484ff. have knowledge because they are *close* to the things – they do not need to rely on rumors – and because they know the *many* things in which the reader is interested: "Quantity, not intensity, is Homer's standard of judgment" (Snell 1946b: 18). A thinker is *multi*-knowledgeable and *multi*-thinking (*poluphrōn, polumētis*), not *deep*-knowledgeable and *deep*-thinking (*bathuphrōn, bathumētēs*) as in the later lyrical literature.

> The character of Odysseus, of the intellectual in the Homeric world, if we may so speak, shows how these forms of Homeric knowledge acquisition interact in the living human being. Odysseus has seen and experienced a lot; he is, furthermore, the *polumechanos* – the one who always knows how to help himself in new ways – and he is the one who listens to his goddess Athena. He did not actually acquire the knowledge that is based on seeing, and the multitude of experiences and insights through his own activity and research; it is more like it happened to him during his many travels. He does not yet qualify as a Solon, of whom Herodotus said that he was the first to go on travels because of his theory, as a result of his research interests. In Odysseus multi-knowledge is oddly separated from his activity in the domain of *epistasthai*. The latter is restricted to finding means to reach a certain goal, in this case to save

73

his and his comrades' lives. And in the third knowledge sphere, that of interpretive understanding, Odysseus, just like all Homeric beings, relies on explicit speech that can be understood: if events, actions, and people do not directly state what they signify, what their meaning is, then a god can do so in intelligible speech; for example, he or she can convey it to a seer, who in turn passes it on to other people, or the Muses can tell it to the poets. Yet neither does the seer research a blank future, nor does the poet attempt to disclose a hidden truth. Rather, the seer teaches us the past, present, and future that a god conveys to him, and the poet reveals only what the muses tell him – and they know everything because they've seen it all and have been present everywhere (as it is said in Bk. II of the *Iliad*). Thus in both cases intellectual superiority rests on the body of knowledge available to the gods in question, and for the Muses Homer explicitly makes it dependent on what they have seen.

(Snell 1962: 48)

Knowledge does not mean moving forward from appearance to an "essence" but accurate placing in relation to the object (process, aggregate) and *adding up* the so-called items of knowledge. It is this that later philosophers would criticize when they said that "Much learning (*polumathie*) does not teach one to have understanding" (DK 22B40).[18] And this is what makes early thinkers' interest in "many wondrous things" (earthquakes, eclipses, the seemingly paradoxical behavior of the Nile), as well as the lack of a connection between the proposed explanations, understandable: geographical descriptions sequentially list the tribes that will be encountered during a voyage or journey, the coastal areas, coastal formations, cities, and characteristics of the inland. They are essentially one-dimensional. Even a thinker such as Thales was content with a list of interesting phenomena and the invention of ingenious explanations for them, without making any attempt to combine these explanations into a system (which would happen only later under Anaximander).[19] The parts into which a situation is dissolved may also include those phenomena

[18] The idea that science consists in the composition of lists can be traced back into the Sumerian past. See Soden (1965), which illustrates and discusses the achievements of the Sumerian-Babylonian list science in a number of disciplines. The difference between Babylonian and Greek mathematics and astronomy consists precisely in this: the first develops methods to describe *phenomena*, and the second develops methods while "let[ting] be the things in the heavens" (Plato, *Republic* 530b; see also *Laws* 818ab). On the Pre-Socratics' critique of the list theory of knowledge see Guthrie (1962: II, 25).

[19] The idea that Thales used water as a principle underlying all appearances, thereby initiating modern philosophy, was first proposed by Aristotle (*Metaphysics* 983b6–12 and 26ff.). However, a closer look at this and other passages as well as of Herodotus' work suggest that Thales actually belonged to an earlier school of thought that would list and explain multiple extraordinary phenomena without combining them in a system. See also the lively account in Fritz Krafft (1971: ch. 3).

74

28. Egyptian drawing of a mortar: Pyramid period

From: Schäfer, Heinrich (1963): *Von ägyptischer Kunst: Eine Grundlage*, Wiesbaden, p. 266, fig. 282.

that we today associate with special observation conditions such as perspective. Thus, for example, in the Old Kingdom (Egypt) vessels were generally drawn with an indentation, thereby turning a phenomenon of perspective into a part of the object itself.[20] In this method of observation the oar that looks bent in the water loses all of the skeptical implications that were later assigned to it.[21] Just as Achilles' sitting failed to give rise to doubts about his swiftness – on the contrary, his swiftness would have to be challenged if it turned out that in principle he wasn't capable of sitting – so does the oar's being bent in the water fail to lead us to doubt its straightness outside of the water – on the contrary, we should challenge its straightness if it turned out that it was not in principle capable of looking bent under water. For the oar that is bent inside the water is not an *aspect* of the real oar, which can exist side by side with its straightness outside of the water as just another aspect. Hence the question arises what the nature of the oar *really* is; and so it turns out that it is nothing but a special *situation* of the real oar, a part of the spatiotemporal complex "oar" that is not only compatible with the latter's straightness in the air but actually required by it.

Even the concept of *law* or *destiny* contains a core that can be

[20] On integrating phenomena of perspective into representations of an object see Schäfer (1963: 266). The red-figure style in Greece temporarily searched for methods to *avoid* any shortenings and other effects of perspective. See Pfuhl (1923: I, 378). Rudiments of the new approach are already indicated here.

[21] The oar that looks bent in the water was used already in antiquity as an argument against the existence of an essence of things; see Sextus Empiricus (*Pyrrhonism* I, 119). In the twentieth century the example has shown up with tiresome frequency in the texts of the sense data philosophers. Example: Ayer (1940: ch. I). The aforementioned Homeric analysis of the oar illusion can be found in Austin (1962). Austin, however, does not consider the result to be Homeric, but generally British. Thus old fairytales keep getting repeated in new clothes, and the authors are not even aware of the repetition.

satisfactorily explained only in an aggregate universe. An aggregate is defined by its parts and their relations to one another. It distinguishes itself from its environment simply by taking up a certain *spatial sector* in it that does not overlap with another aggregate's or another family of aggregates' sector. Now the notion of a sector is included in the concept of *moira*, which presupposes a spatial division of the world into different domains. Such a division is older than the gods,[22] it is superior to the gods and has an ethical dimension, though it does not rest on individual intentions or on an individual purpose. *Moira* is not a person. This is made very clear in *Iliad* XV, where Poseidon defends himself against Zeus' infringements:

> For three brethren are we, begotten of Cronos, and born of Rhea, – Zeus, and myself, and the third is Hades, that is lord of the dead below. And in three-fold wise are all things divided, and unto each hath been apportioned his own domain. I verily, when the lots were shaken, won for my portion the grey sea to be my habitation for ever, and Hades won the murky darkness, while Zeus won the broad heaven amid the air and the clouds; but the earth and high Olympus remain yet common to us all. Wherefore will I not in any wise walk after the will of Zeus; nay in quiet *let him abide in his third portion (menetō tritatēi eni moirēi)*.
>
> (*Iliad* XV, 186–95)

It should be noted here that the gods do not themselves separate the various sectors from one another. What holds them together or separates them is not the gods' power at all. On the contrary – the gods obtain their respective positions and their privileges from the sector in which they reign, not the other way around. It is only later that a personal god's *decision* replaces the *fact* of the world's subdivision, and it is ultimately replaced by the effects of a *spirit* that holds together the various sectors from within. We encounter this purely extensive aspect also in the concept of law. The Greeks had a very keen sense of the connection between *nomos* (law) and *dianemō* (to divide up), while *nomeus* (sector) is a word that denotes the pasture ground assigned to a particular shepherd.[23]

---

[22] In *Iliad* XIV, 200f., Oceanus is the originator of the gods; but in Hesiod Chaos, Gaia, Eros, and then Uranus, the dry Earth and Pontus, which were given birth to by Gaia "without sweet union of love" (*Theogony* 132) – that is, without mediation of personal principles – are above the personal gods.

[23] On these remarks regarding *moira* and *nomos* see Cornford (1912: ch. 1). Cornford also offers an interesting explanation of the origins of these concepts, which, however, today no longer can be regarded as fully convincing: "*Moira* came to be supreme in nature over all the subordinate wills of men and Gods, because she was first supreme in human society, which was continuous with Nature. Here, too, we find the ultimate reason why destiny is moral: she defines the limits of mores, of social custom" (Cornford 1912: 51).

To sum up: both the formal structure of the epics and the content of the concepts used in them, both ideology and everyday thought, both the structure inherent in language and the utterances formulated in it, both the experience of self of the Homeric person and the critique of subsequent philosophers, and both the concepts of knowledge and the recognized examples of knowledge, show that perception, *thought*, action concurrently repeat those elements that we could *clearly see* in the late geometric and even in the archaic style. These elements are therefore not merely a spawn of artistic imagination or poetic exuberance. They are elements of a *world* in which the Homeric person as such lived and which was for them as *real* as our own world is *real* for us today. This clearly refutes naïve naturalism. *Western philosophy of nature*, however, is generated in the transition from this world to the very differently structured substance world of the Pre-Socratics.

3.3. Views of Reality and the Language of Science: Some Basic Considerations

The concept of *reality*, which we often use in order to reveal views different from our own as "an illusory representation of man and the world, an inaccurate explanation of the order of things" (Godelier 1971: 96), can be used *externally* or *internally*. The external approach compares a theory or point of view with a widespread and popular ideology such as science, and recognizes only those components as "correct" or "real" that cohere with the ideology. The rest is error, illusion, or collective imagination. Myth, as Godelier points out accordingly, turns the "numerous objective data" about nature "into an 'imaginary' explanation of reality" (Godelier 1971: 101), with "objective data" meaning the data of science, which also gives us "the *raisons d'être* and necessity of the multiple motion of history which [. . .] itself remains essentially the same, with new contents to think" (Godelier 1971: 110; see also Althusser 1965). The internal approach asks how people who use certain artistic, linguistic, or conceptual media can distinguish between reality and appearance, and how their selected distinction is reflected in their perceptions. It also looks at the ideological effects on the reality thus constituted. The external view obviously relies on the internal one, for it is the internal view of a *particular* form of life combined with the naïve belief that this form of life is the only correct one. Nietzsche decidedly employed the internal view in his examination of Greek mythology:

77

Indeed, it is only by means of the rigid and regular web of concepts that the waking man clearly sees that he is awake; and it is precisely because of this that he sometimes thinks that he must be dreaming when this web of concepts is torn by art. Pascal is right in maintaining that if the same dream came to us every night we would be just as occupied with it as we are with the things that we see every day. "If a workman were sure to dream for twelve straight hours every night that he was king," said Pascal, "I believe that he would be just as happy as a king who dreamt for twelve hours every night that he was a workman." In fact, because of the way that myth takes it for granted that miracles are always happening, the waking life of a mythically inspired people – the ancient Greeks, for instance – more closely resembles a dream than it does the waking world of a scientifically disenchanted thinker. When every tree can suddenly speak as a nymph, when a god in the shape of a bull can drag away maidens, when even the goddess Athena herself is suddenly seen in the company of Peisastratus driving through the market place of Athens with a beautiful team of horses – and this is what the honest Athenian believed – then, as in a dream, anything is possible at each moment, and all of nature swarms around man as if it were nothing but a masquerade of the gods, who were merely amusing themselves by deceiving men in all these shapes.

(Nietzsche 1873b: 23f.)

The concept of reality, which impacts people's actions, feelings, and perceptions within a certain period of time, thus depends in part on psychological, social, and cultural factors. The factors change partly under the pressure of external circumstances, and partly as a result of conscious human efforts (rationalists tend to grossly overestimate the second part). Their impact is more or less "rational," and "reason" itself is determined either by these factors themselves or by the factors of a competing ideology. "Absolute" standards, that is, standards independent of all factors, may exist in Heaven, but here on Earth with its diverse forms of reality there are none. Thus the formal structure of the epics, which has a decisive impact on the content and the perceived world, is "rational" in our sense, if we take into account the requirements of oral poetry and tradition. It is not "rational" in our sense with regard to the purpose of representing "the world" – and with that, of course, we mean *our* world – adequately. It is rational for the epic poets, who would not get very far without it, and it is furthermore rational for their audience, which *sees*, *feels*, and *depicts* the world as an aggregate composed of invariable event atoms. However, a *conscious* selection of rationality criteria is always random in the sense that the underlying ideology is simply taken for granted without much argument. If the ideology is popular then all efforts are focused on it, success rates increase, difficulties are either removed or presented with great optimism as just temporary, and it seems that there are "objective" reasons to give it preference, since all success is but the result of better performance (this is, by the way, also how the rise of modern science can be explained).

These important and fairly simple considerations, which reveal an interesting ambiguity in the concept of reality, were rarely taken seriously, and under the pressure of modern science everything was readily brushed aside as appearance that did not confirm to *its* rules. Even such progressive thinkers as de Santillana, Lévi-Strauss, and – to venture into the abyss of politics – Althusser did not think for a second that they should put principles of form belonging to "primitive" thought *on an equal footing* with scientific principles. Even less were they willing to occasionally give them *preference*. The most that they were willing to concede was a watered-down version of the theory of double truth: "There is only one solution to the paradox [of comprehensive Neolithic knowledge], namely, that there are two distinct modes of scientific thought" (Lévi-Strauss 1962a: 15). Since myths do not allow for meaninglessness (Lévi-Strauss 1962a: 22), they can affect science in a "liberating" way, and in this they are actually superior to the latter. And yet a closer comparison of different cosmologies almost always reveals the existence of advantages *and* disadvantages on both sides. This will become obvious in our discussion of the various historical transitions.

Let us take another look at the features of Homeric epics discussed in the previous section. In a first (and for the purpose of this essay entirely sufficient) approximation, the epics consist of two *layers*. The first layer is the *sequence of events* that comes immediately to our attention while we read or listen – the episodes of the Trojan War and Odysseus' adventures. Not all of these events are pure legend, and some important discoveries rest on the assumption of their historical truth (Schliemann, Evans, Nilsson). Still today reports in the epics as well as in the accompanying legends are used in research and for bridging or at least somehow explaining gaps in history.[24] And yet in this first layer we find truth and falsehood, history and legend thoroughly intermingled. The second layer is the *structure of the event sequences*, that is, the type of objects and events described as well as the form of their interconnection. This structure leads us beyond individual circumstances and reveals *general* features of human beings,

[24] As an example see ch. 2 of Whitman (1958).

their social world, their physical world, and of the world of gods. The contents of this second layer are essentially *cosmological*.

The dualism of event sequence and structure is implicit in every intelligible narrative, be it a factual report, a fairytale, a myth, or a plain tall tale. Every narrative consists of elements that *are used in accordance with certain general rules* and in this way *express* or "constitute" *general ideas*, whereby their impact can be noticed even in some very simple truths. Even a historian striving for entirely singular facts relies on general rules and on the constituting processes derived from them. They are required for people to understand historical accounts and be able to pass them on. Even just the fact that historians use a script consisting of letters, and that they put down their contents on paper in a linear order, turns the mass of events into series of events that, while brought into the shape of a long tapeworm, will never reach the level of mutual enmeshment realized, for example, in a musical score, in a three-dimensional model, in a movie, or on the stage.[25] The order conveyed in the grammatical structure of the language is superimposed on the linear order: in the sentences events commonly appear as "coincidences" that happen to certain subjects, or as "actions" generated by them. As historians point out, it is not that "war happens," or even that the "king's war happens," but rather, "The king wages war." Thus we have not merely linear processes but also intersections, short dramas with a beginning and an ending, and power nodes sending out unilateral radiations.[26]

The double nature of every intelligible story is responsible for the fact that a report can be misleading or "false" in either of two respects. It can be misleading because the singular sentences it uses are false – the events are not accurately reported – or it can be misleading because it approaches the world with unusable categories, because it introduces regularities, situations, relations that do not exist. For example, the sentence "Achilles dreams" can deviate from the truth in two respects: first, because Achilles did not actually dream at the time indicated; and second, because dreaming is not an activity performed by a subject – though the phrase "to dream," as it is used today, suggests this – but an external intervention of the kind described in section 16 [*chapter 3.2*]. In the latter case it is just as inappropriate

---

[25] See Lévi-Strauss' remarks in the introduction to *The Raw and the Cooked*, where he explains his own, more symphonic type of narrative.

[26] Marshal McLuhan has colorfully explored the history and impact of the linear method of representation, especially in McLuhan (1965). On the impact of the narrative style on the contents of historical accounts see Hughes (1964: ch. IV).

to say "Achilles dreams" as it is to say "Achilles blushes" when he is being coated in red paint.

Note that speakers of a given language can only rely on the forms of their language while attempting to explore the adequacy of certain phrases. When separating "language" and "reality" in order to measure the former by means of the latter, Homeric singers have to drop the separation again as soon as they start describing "reality." For they describe reality through the categories of their language, and that's also how they see it. This apparent impossibility of a fundamental critique of means of expression (if we use another language then the argument essentially remains the same) has caused certain thinkers to look for languages that precisely describe *what is given*, not more or less. A look at the languages proposed shows that they do not manage to solve the problem. Bacon and his contemporaries used a condensed, staccato-type language without "poetic" elements. That is, they assumed that the world consists of *simple things lying side by side* and not surrounded by any baroque aura: the dry language reflects the new dry mores and the dry manner in which the world is perceived by dry people. Yet they did not come any closer to "nature" that way.

In general: a series of "false" singular sentences describing *episodes* that did not take place can nonetheless accurately reflect the *general features* of the surrounding world, *and usually does reflect them accurately* – for the general features of the series, and the general features accompanying its application, and the sensations, actions, emotions accompanying its application first provide us with the criteria needed to distinguish reality from appearance. Thus such a sentence mass introduces certain cosmological assumptions. Though the assumptions are not explicitly formulated, they are *insinuated* and are therefore not easily accessible in direct examination.

In the case of the epics the world consists of a number of close-knit series of events each composed of relatively short basic events. Both gods and humans are responsible for events' occurring; there are no coincidences. The nature of the agents and of the things they move is not something hidden that underlies the actions and their effects, provides them with a uniform substratum, and has to be revealed through special methods (ecstatic states and so on). Even that which humans cannot attain due to their limitations can be described in sentences that are clear and intelligible to them. Nature does not *appear* in the events; rather, it is *taken apart* in them, and fully so. Just like a casern consists of houses and barracks that do not need to be supplemented by yet another special entity in order to become a casern, so does an object consist of its components and its behaviors

81

under certain conditions, *and only of those*. Complexes can permeate one another, as, for example, when a divine act of interference triggers anger in a human being, or when a dream appears to the dreamer.

The procedure of *taking apart* an object becomes especially clear in late geometric art. Here the elements are not aspects – giving the observer the *illusion* of a reality different from its aspects – but part of diagrams, that is, of *visible catalogs*[27] representing how the object is composed of elements in certain ways regulated by conventions. We may assume that this convention, which was linked to the needs of royal courts as well as to mnemonics, leads to a special view of the world, eventually finding support even in actual *experience*. This assumption is confirmed both by conceptual analysis and by later philosophers' criticism: humans subjectively feel the contribution of external factors in their decisions, emotions, ideas, and their entire inner life is *open* in the sense that agents other than the person herself can participate in it; it is not a closed power node or an "I." Instead, it is a *way station* subject to factors, events, and coincidences. Objectively we see others as mannequins who are involved in various actions and whose nature is *revealed* in these actions. Even inanimate objects do not show more or less deceptive aspects (such as a straight oar and an oar bent in water), but behave differently in different circumstances; they have their characteristics *due to* these behaviors and not in spite of them. Truncations, indications of perspective that might suggest another, substance-filled view, were suppressed and remained unnoticed, just as even today we barely notice (physiologically rather strong) afterimages. Thus the cosmological skeleton of the epics corresponded to the facts of experience, was supported by them, and contributed in turn to a more precise articulation of them. The skeleton and the cosmology articulated through it *are true*, because they provide a correct representation of the world in which Homeric people lived.

We encounter the same phenomenon of a correspondence between means of expression, ideology, and "objectively experienced" world at the beginning of the modern sciences. Just like the Homeric ideology, so also is this primordial scientific worldview confirmed by "objectively ascertainable" experiences. And just like Homer it can be "critiqued" and ridiculed from the *outside*, that is, by comparison with other viewpoints naïvely considered correct. Let us look into how such a critique might proceed. The ideology of the seventeenth

[27] See again Soden (1965) on the Babylonians' catalog theory.

century emphasized objectivity, dryness, and lack of exuberance. There was "a tendency toward increasing interest in prose during the course of the century," wrote Robert K. Merton.

> This trend is not totally unrelated, as we shall see, with the similar development of interest in science: both fields were concerned with the exposition and description of empirical phenomena. The emphasis [. . .] was on the descriptive and "true" rather than on the imaginative and fictitious. [. . .] The scientific standard of impersonal denotation was applied to all forms of literature, which traditionally is personal and connotative.
>
> (Merton 1938: 19f.)

In accordance with this John Dryden emphasized in his *Religio Laici*: "And this unpolished rugged verse I chose as fittest for discourse, and nearest Prose" (quoted from Budick 1970: 20). This context is also the background for the following quote from *The History of the Royal Society*. The Royal Society, wrote Sprat, had decided

> to reject all the amplifications, digressions, and swellings of style; to return back to the primitive purity, and shortness, when men deliver'd so many *things*, almost in an equal number of *words*. They have exacted from all their members a close, naked, natural way of speaking; positive expressions; clear senses; a native easiness: bringing all things as near the Mathematical plainness, as they can.
>
> (Sprat 1667: 113)

The quote, whose ideology underlies the efforts of "scientific" circles to drain and butcher human language, contains a number of hypotheses all of which are false when compared to our *contemporary* approach. The *first hypothesis* is: a "close, naked, and natural" way of speaking, whose meaning is always clear, is close to mathematics, which possesses maximum clarity, brevity, and naturalness. This may be true for certain branches of mathematics (though it would always be wrong to say that mathematics has a "natural" way of speaking), and for certain forms of *representation* in these branches. It is true especially for those domains that are temporarily in a state of stagnancy (as was Euclidean geometry at the time of the quote) and that are therefore terminologically firm. In such cases the "close, naked" claim really matches the facts. But it is different for those domains in which progress is made in a rapid succession of steps.[28] The lack of clarity in seventeenth-century mathematics, especially in Newton's new mathematics, was indeed

---

[28] See Pólya (1954) and especially Imre Lakatos (1963/4).

soon noticed, e.g., by Berkeley in his critique of the fluxionary calculus.

The *second hypothesis* claims that language in early, "primitive" times was simple and short, and that the number of words exactly matched the number of things. This second hypothesis has already been refuted by recent research.[29] Though the myths of the earlier times had an astonishingly rich factual core of astronomical, biological, physiological, and social facts, this core was surrounded by "amplifications, digressions, and swellings of style"; it was embedded in an artistic and religious context livelier than today's art and religion, and it was often accompanied by ritual acts.

According to the *third hypothesis*, a style is acceptable only if the number of things exactly matches the number of words. This hypothesis implies the cosmological assumption that the world consists of things that have a relatively isolated existence, as well as the semantic assumption that language should precisely reproduce this world. The world is not a mechanism but a mechanical aggregate. This cosmological assumption seems very questionable today. The interaction between an experiment and its environment, and eventually the entire (finite) world, make it necessary to represent the latter by a single wave function, and this wave function cannot be split up into more basic wave functions that would denote individual things and their interactions (though approximations are of course possible). Furthermore, the non-linear character of general relativity theory does not allow us to simply add new forces to a given situation – the entire space-time continuum would have to be recalculated. The semantic assumption is not acceptable, since languages have limitations that do not show up in empirical evidence but only in comparison with other languages.

According to the *fourth hypothesis*, which is at the basis of this short and almost trivial quote, exuberance, "amplifications, digressions, and swellings of style" are never bound to the subject matter. This means: the *world is a bleak, non-baroque world*. There are no swellings in it. This may be true or not true – in any case, the assumption is introduced without any supporting argument. It is soon expanded to cover human beings as well (behaviorism), thereby changing them so drastically that they eventually conform to it in every detail. The dry and colorless ideology of the inventors of modern science created a

---

[29] See sections [4] [*chapter 1.2*] and [5] [*chapter 1*] as well as [9] [*chapter 2.1*] above.

dry and colorless world and eventually also dry, colorless, and almost no longer human people.[30]

The *fifth hypothesis* that we can extract from the short quote assumes that the categories of language can be confronted with a world that has not yet been linguistically grasped ("men deliver'd so many things almost in an equal number of words"), and that the confrontation decides whether or not the language is adequate. Now since humans cannot argue without language, their original grasp of the world cannot be but a *primordial intuition*. So we start with this primordial intuition and proceed to gradually build up a language that fits it. This hypothesis of the origins and testing of languages has been refuted by linguistic, ethological, psychological, and logical considerations. Every living creature has a set of built-in schemata even at birth, which enable it to find its way around in the world.[31] Without such schemata neither life nor thought would be possible. Even inanimate matter reacts to its environment selectively, thereby revealing its specific nature. The question is not how to get around *without* categories, but how best to select one's categories and whether there are categories concerning which we have no choice. We can see here how even the allegedly so "natural" and "unbiased" approach of the

---

[30] Mailer (1970: ch. 2, "The Psychology of Astronauts").

[31] See Lorenz (1935b) as well as Piaget's and Chomsky's work. The need for categories in the composition of a world becomes especially clear from the failure of the naturalists' attempts to represent life "as it really is." According to Heinrich Hart, Arno Holz "developed his view based on the example of a leaf falling from a tree. The old art did not know more about the falling leaf than that it swirls to the ground. The new art describes this process from one second to the next; it describes how the leaf now in this second gets illuminated from this side, lightening up in a reddish color, while simultaneously appearing shaded gray on the other side, and how this gets turned around in the next second. It describes how the leaf first falls vertically, then gets driven on its side, and then again sinks vertically; it describes – well, heaven knows what else it has to report" (Hart 1907: 68f.). Indeed, only Heaven knows, and it is also clear that the writer Arno Holz implies this heavenly knowledge of the phenomena's *infinity*, their partial indeterminateness, and their lapse into the subjective realm (Arno Holz' language is always clear and precise, it captures precise boundaries of elements within a continuous process, and composes the latter out of the former like a mosaic). And yet it is also clear that he either does not know this infinity or has not fully grasped it. The description becomes more detailed, the view more microscopic, and the categories adjust to this microscopic view, but that is all. That a microscopic description can be dreamlike and illusionary despite its apparent factual nature can be seen in Homer. Homer's battle descriptions are more detailed than battle descriptions in other epics. We find precise characterizations of the movements of wounded warriors comparing them to lifeless objects, there is great detail in the descriptions of individual battles, whereas elsewhere we often find only vague indications of large commotions. And yet Homer's precision often leads to completely illusionary results. "Almost every traveller who examines the plain of Troy or the isle of Ithaca identifies the various landmarks to his own satisfaction, but the identifications are all different" (Murray 1934: 160).

early empiricists introduced a number of cosmological, historical, and physiological assumptions of exactly the same kind that we have found in the epics. The only difference is that the assumptions are now explicit and hence more easily accessible to examination.

— 4 —

TRANSITION TO AN EXPLICITLY CONCEPTUAL APPROACH TO NATURE

[17] In the seventh and eighth centuries Homer's colorful and detailed image of humanity and nature underwent a series of changes that gave rise to Western *literature, philosophy, and science.* From the early myths and their complicated genealogical systems, from the Homeric formalization, clarification, and refinement of their content, and from Hesiod's earnest reshaping of the product of those processes, the development proceeds on different routes with different sequences of thought and imagery, thus leading in different directions depending on temperament, social situation, language, geographic location, and random historical events. Both natural changes and deliberate intervention create new institutions such as the hoplite army, the genre of tragedy, international trade, scientific arguments, philosophical schools, and political parties, until in the fifth century BC in Athens we eventually find ourselves confronting science, politics, literature, philosophy, and numerous other disciplines as *independent* and in many respects modified. It is very tempting to replace this complex network of relations with a few simple connections, and we are easily rushed into highlighting the apparently familiar and the seemingly modern as a commendable first step toward a more "rational" view. Thus in retrospect everything becomes more simplistic, and we do not learn anything about the manifold and often very surprising *coincidences*, without which philosophy and the sciences would not have been possible in the first place.

The principle of simplification at work here is always the same: whatever corresponds to reason and perception deserves mentioning and receives detailed descriptions, and attempts are made to show what rational steps led to its discovery (it is usually taken for granted that rational things can come about only in a rational manner). The *rest* of the story is explained away either in terms of 'factors' or based on a (frequently very naïve) psychology of error. Two examples may illustrate the procedure. Nobody makes a fuss over the fact that the spots on the moon are mentioned in numerous astronomical and astrological texts. For the spots on the moon are 'real'; everybody can see them, so it is not surprising that they were also seen in the past. But reports about the interference of gods in earthly matters are a different case. Since the gods do not exist, there is nothing that could be seen, and hence we have a problem: the reports need "explaining."[32] The second example illustrates our attitude toward the Pre-Socratics. Everyone praises Anaximander for eliminating the personal gods – for being the first one who eventually approaches nature through sober observation, as befits a rational human being, and for removing the phantoms of an insufficiently tamed imagination. However, his theory of dualities, in which heat includes the heat of love and coldness the chills of hate, is thrown aside as a sad leftover of past confusions.[33]

The popular assumption that ancient ideas consist of a rational core embedded in an irrational padding dates back to Hecataeus of Miletus.[34] Even an author as insightful as Kurt von Fritz shared it. According to Fritz the Pre-Socratics eliminated the gods *because they were not part of our everyday experience*: "The criterion of Anaximander's critique is whether something contradicts the general nature of directly accessible experience, and apparently Anaximander applied this criterion also to things such as the distant past or the distant stars that are not accessible to direct examination and observation" (Fritz 1967: 47). Similarly Fritz Krafft, in his otherwise excellent introduction, traces back the elimination of the gods to a strict application of Anaximander's principle *opsis gar ton adēlon ta phainomena* (DK 59B21a) as early as in Anaximander (Krafft 1971: 126).[35]

---

[32] Yet we do not need to mobilize the gods in order to become convinced that our perception of the environment has not been the same at all times. For example, we have good reasons to assume that the moon and the stars were not only differently *described* at different times but also differently *seen*. See the problem and material discussed in ch. 10 of my essay *Against Method* (Feyerabend 1975).

[33] Regarding the Pre-Socratic qualities hot–cold, light–heavy, dry–wet see Fritz (1946: 19f.) as well as the article on Protagoras in Wissowa's *Realenzyklopädie* (1958: vol. 45, col. 914). "Physical heat and the heat of love really are the same quality, as are physical coldness and the coldness of hate. For the same reason the wetness of a 'moist soul' in Heraclitus is obviously not just a physical quality but also a character quality the appearance or feel of which does indeed trigger sensations of the same type as those triggered by touching a moist object" (Fritz 1967: 45).

[34] See above, section [8] [*chapter 2.0*].

[35] [*Editors' note: Feyerabend did not translate the Greek text: "opsis gar ton adēlon ta phainomena" – "appearances are the face of the hidden."*]

However, Anaximander could not have used this principle since, first of all, his explanation of the earth's central position as a position of rest due to symmetry (DK 12A26) is incompatible with "the general nature of directly accessible experience." Second, as we have seen, the experience of divine intervention was part of the everyday experience of people in the Homeric age (and probably their predecessors as well). If the elimination of the gods had been based on the principle mentioned by Fritz then it would have had to be preceded by a change in the "general nature of directly accessible experience" itself, which in turn cannot be traced back to the invention of rational principles but rather is based on more comprehensive reasons and causes. Skipping this change means suppressing an essential element in the development of philosophy of nature and trivializing that development: it means transforming it from a social process with enigmatic causes, comprehensive effects, and undesired side effects into an invention by some specialists that subsequently continues to be revised and polished by an aggressive and power-hungry club of like-minded fellows.[36]

Of course, these assessments are based on naïve Naturalism:[37] if people have the same experience at all times and are able to take advantage of the same faculty of reasoning then fundamental deviations from the present point of view are indeed nothing but the result of oversights and lack of intellectual discipline. They are nothing but interesting (and somewhat depressing) *psychological* processes. They have nothing to do with the development of knowledge. However, naïve Naturalism is false.[38] There are forms of life fundamentally

---

[36] This is the idea that Popper proposed in his essay "Back to the Presocratics" (Popper 1958); it goes back to Aristotle (*Metaphysics* 981b11ff.).

[37] See above, section 10 [*chapter 2.1*].

[38] The shortcomings of naïve Naturalism are especially striking in the case of astrology. Compare, for example, Robert Eisler's masterpiece *Weltenmantel und Himmelszelt* (1910), which is based on diligent research into sources, traditions, works of art, with his ranting pamphlet *The Royal Art of Astrology* (London 1946), which plainly denies early man all reason, observation, and inventive talent and sees as the task of science the establishment of not-too-bold generalizations based on experiences as comprehensive as possible: "'Prescientific' observers have a natural tendency to reach overly rash generalizations on the basis of limited experience" (Eisler 1946: 140). History itself, however, has refuted this. In Homer we encounter but few generalizations, and whatever appear to be imaginative generalizations – such as the assumption of divine intervention – are in reality faithful renderings of direct experience. Science, however, started out with assumptions such as that the celestial bodies, including the planets, have a firm orbit or that the Earth floats in an empty space without any support. Such assumptions far exceed familiar facts and often contradict them. It is obvious that a view like this (which we also encounter in other authors such as Boll, Gundel, and Cumont) necessarily leads to a distorted notion of the development of philosophy of nature as well as of science.

different from ours. Homer's worldview combines concepts, artistic style, a formula system for epics, ideology, and "theory of knowledge" (which is not yet a *theory*, that is, a structure built on explicit principles) in a manner – and people follow them, adding a kind of certainty – that can be explained only in terms of an *experience of the world* different from our own. *An animated world, divine intervention, the "openness" of the soul's life are not preconceptions or errors or results of a superficial approach, but clearly recognizable components of this experience of the world, and their elimination constitutes an elimination of important knowledge.* Moreover, the elimination of the Homeric worldview *as a whole* eradicates not merely one or other *piece of knowledge* but the foundations of *all* knowledge that can be gathered *in principle* (that is, with the help of the clear- and far-sighted Muses) in the Homeric world. A whole world dissolves, including its human observer, and is replaced by another world with another human observer. We cannot expect that the events that take place during such a change of worlds will observe a "rational" rule.

This last remark applies especially to the development of the concepts. Neither conceptual analyses nor methodological hypotheses, nor even the simplest rules of logic, can teach us how the pieces of knowledge emerging from the broken-down structure were once hooked together and how they gradually combine to form a new conceptual structure. Separated from strict cognitive relations, the ideas are now subject to diverse non-conceptual factors, *and only empirical investigations* can determine which of these factors leave traces and what the nature of these traces is.

At the same time it does not suffice to register solely the historical succession of ideas and their dependence on causal factors. Even those concepts that enter the stage of history in a most irrational manner, and that for a long time undergo not logical changes but merely external ones, still have a tendency toward reciprocal adaptation as well as of adaptation to the material, part of which is to be described and another part to be constructed. Logical relations are never completely absent, and the attempt to find them is never completely unsuccessful. Thus the correct historical account is the one that combines analysis of logical relations with a description of external and non-logical factors and shows when and why one or the other side comes to the fore. We rarely encounter such a "mixed" account in the literature. Nor can one be satisfactorily provided in the present short introduction. And yet we can *prepare* the way for it by extending our limits a little and at least *mentioning* some unexpected and surprising events. This is the reason why I discuss visual arts, tragedy and events in

90

the history of wars side by side with more philosophical material. During the time to be described here there was not yet any institution "philosophy of nature" that would protect our thinking from non-philosophical factors and thus ensure the purity of its notions.

The impression of a state of chaos that such an account will inevitably generate, however, is only in part the result of my efforts to achieve brevity *and* completeness. Like any historical event, the birth of the world of philosophy depends on numerous relatively independent factors the random coincidence of which, only slightly altered here and there by human reason, generates new structures. A loose list of development chains and their interaction, which are supplemented in various ways and whose elements are combined in diverse manners, is more appropriate for this situation than a strict "systematic" construction that eventually serves to support once again the myth of the *one* faculty of reason and the *one* experience of the world.

4.1. The New World of the Philosophers: Advantages and Disadvantages

[18] Let me then start with a list of some *development chains* that contributed to the emergence of Western philosophy of nature. Indirect descriptions of cosmological principles (in which I always also include notions of human nature) by means of suitably structured myths are replaced by their explicit formulation, first through a picturesque *cosmogony* (Hesiod, Anaximander) and subsequently through the imageless, gray medium of *concepts*. The *gradual growth* of notions and their natural adaptation to the demands of their environment are replaced by their *conscious transformation* in the course of a debate between "rational" people. And we take full advantage of the liberty that we now have with regard to our concepts. We playfully introduce, use, transform and abandon countless thought patterns. We invent the hypothesis, the fiction. Both liberate literature and philosophy from the constraint of truth *reports* and enlist them in the service of *establishing* truth and opening up *possibilities*. The highlight of this experimental phase is the Sophistic, which shows us humans how much depends on our *decisions* and *inventions*, thereby at the same time using this dependence to dissolve traditional laws, morals and rules. But soon our process of thinking is once again subjected to more stringent rules – the origins of logic by detour of dialectics – and thus gradually emerges the *systematic treatise*, which is no longer dependent on random coincidences of dialogue but determined by the

"inner logic" of ideas: though as humans we become the masters of our thoughts, we at once abandon our sovereignty and relinquish it to the logicians and methodologists, the natural successors of a priesthood concerned for the purity of fundamental articles of faith.

Not just the form but also the *contents of thought* undergo an essential change. The Homeric world consisted of closely tied parts clustered into aggregates, while the "new world" of the philosophers consists of relatively isolated *substantial units* whose nature can be only roughly guessed on the basis of deceptive appearances. (Innate schemata of guessing are used to transform drawings from visible lists of objective parts into requirements of the illusive perception of superior entities. Instead of a sign *reminiscent* of a part or object, just like a word may be reminiscent of it, we now have *stimuli pretending to be* the object.) The *Homeric human being* was an open aggregate of limbs, feelings, and perceptions, a playground of partly internal and partly external elements of consciousness, a switchboard processing countless factors. The *new human* is an "autonomous subject" with self-willed ideas, motives, and feelings, and this subject is both emotionally and epistemologically separate from its environment (now only humans have feelings, while nature is insentient). A uniform system still entirely uncomprehended in its uniformity – the human being – faces a number of uniform substances equally uncomprehended in their uniformity. Conceptually this is reflected in a novel and equally uncomprehended notion of *knowledge*: in lieu of a manifold of concrete mediations between micro-aggregate human and macro-aggregate world, giving rise to as many concrete relations (as many kinds of concrete knowledge), there is an abstract notion no longer connected with any concrete circumstances and thus at first entirely *void*. Not only is the notion of knowledge itself void, but in addition, the knowledge expressed in terms of this notion has lost much of its former content. Everywhere "words [start] to lose their content and become formulaic, void, unidirectional" (Fritz 1938: 11). Especially in philosophy, the Greek language in the fifth century enters a state in which "the more refined and precise distinctions of the older language are often blurred" (Fritz 1971: 78).[39] Concentration on abstract ideas of little content further leads to an obliteration of our community with the world and other humans. We notice "an increasing alienation. For this certainty of clear correlations destroyed the meaning of

---

[39] Our contemporary philosophers of ordinary language, especially Austin and his unfortunately not very plentiful successors, struggle against a similar obliteration of the English language through the concepts of an abstract (Hegelian, positivist) philosophy.

the world [in a region separate from the world of appearance] as it had been expressed in myths or poetic conceptions [even still in the colorful science of the early Ionians]" (Snell 1924: 80). Parmenides represents the high point of this destruction. The philosophy that systematically developed the novel, more abstract notions isolated itself from the life of tribe and city; it became a specialty practiced by the learned, who later would transform the city's life from the outside and often by means of force (Plato in Sicily!).

Like every historical process the separation of humans and their environment, which was anticipated in all these developments and whose effects could hardly go unnoticed and be mastered even today, has advantages and disadvantages. The separation *liberated humans*, it transformed them from a *component* of nature and society, directly subjected to the impact of both, into their *observer* and *transformer*. No longer was the world simply there; rather, it became something alien that had to be conquered anew, both conceptually and practically. The new viewpoint became very obvious in the new concept of time, which turned from an approaching stream into a thing "that is in our own actions and subsistence" (Fränkel 1960: 14).[40] Even human existence was no longer simply a given; it had to master its new position both conceptually and emotionally in order to meet the challenge of isolated things versus fellow humans. Conceptually this led to the experimental phase already mentioned, which in turn opened up into the new bond of a gradually developing *school philosophy*. New institutions such as democracy, which lends the individual a hitherto unknown latitude of action and thought, tragedy, which indirectly and by use of concrete emotions proposes new behaviors and critiques old ones, religious community, which attempts to reinstate the old tribal morale, and philosophical schools, which attempt similar things in a more conceptual manner, served to solve the emotional and social problems. Oriental ideas such as the immortality of the soul, which does not have any place in Homer's world, were used successfully to reunite the individual with its world.

The opportunities created by the aforementioned separation were not always properly utilized. Transitory states within the development process were often *preserved* instead of serving as desirable starting points of a new and more elastic unity between humans as well as between humanity and nature, thereby enabling a reanimation

---

[40] The Nuer's conception of time fully corresponds to this ancient notion; see Evans-Pritchard (1940).

of nature and a revival of humanity. Moreover, the ideology increasingly incorporates a certain *lack of tolerance*: the ancient acceptance of diverse ideologies, which resulted from an understanding of the diversity of both world and humanity, is replaced by *exclusiveness* and harsh judgment from self-righteous deities now enlarged into abstract powers – and hence *dreaded*. And those entities are not merely a philosophical idea. They become *visible* in the esthetically pleasing yet totalitarian *harmony* of classic sculptures. They can be represented most realistically by the use of *perspective*, which in contrast to the previous aggregate way of perceiving things presupposes and teaches "viewing nature contiguously" ("Zusammensehen der Natur," Schäfer 1963: 277), and whose artful articulation is paralleled in the developing articulation of Greek sentence structures with their delicate system of subordinate and coordinate clauses. Furthermore, they are *felt* and described in *lyric poetry*, and they are *constituted* and critically evaluated in *drama*. Frequently the result of all of these developments in this increasingly independent domain of thought is the replacement of a complex, recognizable, and for the most part already recognized social environment, rich in colors and emotions, with some simple, abstract, clear, yet colorless principles (of natural movement, of human behavior) that are alien to us humans and discernible only in a corresponding colorless reality by means of a matching colorless kind of perception. Only in such an environment can Anaxagoras' principle *opsis adēlon ta phainomena* (DK 59B212) lead to materialism. From now on it is this interplay of stabilizing (impoverishing, lifeless-making) and brightening (enriching, stimulating) tendencies that guides our thinking. Clearly, phenomenal and conceptual changes such as the ones just indicated do not take place smoothly. As an example, let us look at one of the logical difficulties that arose during the transition and which is obviously applicable in the consideration of nature.

In the *Iliad* IX, 308ff. Achilles tries to articulate the idea that the gifts Odysseus offers to him on behalf of Agamemnon do not manage to restore his honor or to eradicate the ingratitude shown to him. Furthermore, he tries to differentiate between honor, gratitude, and respect, on the one hand, and their visible social manifestations, on the other. We might be tempted to say that he discovered the *true meaning* of honor, that he discovered honor to consist of more than just being honored: "Coward and brave man both get equal honor [in the sense of being honored]" (*Iliad* IX, 319). But this interpretation overlooks the fact that Homer's language – which is, after all, the language Achilles uses – does not have any

94

means to express a "discovery" of this kind. In Homer the nature of an object is fully determined by a list of its components and their relations. The "components" of honor that Achilles lists in *Iliad* IX, 225ff. fully *constitute* honor, gratitude, and respect. To deny honor where these components are all present is not a discovery but a misuse of language. Achilles, according to Adam Parry, "has no language with which to express his disillusionment. Yet he expresses it, and in a remarkable way. He does it by misusing the language he disposes of. He asks questions that cannot be answered and makes demands that cannot be met" (Parry 1964: 53). Once such *misuse* of language runs rampant and is regularly practiced, it turns into the preparation of a new language that distinguishes between the *nature* of an object and its social and perceptual *manifestations*. The new *language*, once it spreads, affects the articulation of our spiritual life and thus of our perception, and we as humans soon find ourselves in a new environment, we perceive new objects, we live in a *new world*.

Now, the creation of new worlds is not an instantaneous event. The conversion of a sophisticated language takes considerable time. During the transition period, the older view still remains dominant and provides our criteria for distinguishing between sense and nonsense. Thus, linguistically the transition is full of riddles and paradoxes. The critique of *polumathie* (Heraclitus in DK 22B40 – *polymathy*) is one such paradox, as is the similar critique of description as a mere enumeration of components in Plato (example: *Euthyphro* 6de), for according to the old view *there is nothing but* components and their arrangement. Also paradoxical is the sentence "The limits of the soul you would not find out, though you should traverse every way, so deep lies its principle" (Heraclitus DK 22B45), for the "soul" in turn consists of nothing but an order of events that can in principle be grasped, enumerated, and clearly isolated from one another. Other paradoxes include Sappho's "bittersweet eros" and Anacreon's "I both love and do not love, am mad and not mad," neither of which has a place in the Homeric human's clearly defined, *physically structured* experience. Especially paradoxical is the distinction between a "true world" and "deceptive appearances" and the identification of *all* regular knowledge with the latter, which renounces the well-articulated arrangement of Homeric aggregates, paving the way for a "chaos of appearances" (whereas the aggregate universe does not acknowledge "appearances" or "chaos" in the first place). *Components* are removed from this arrangement and demoted to *appearances* that can provide us only with uncertain indications

95

of the "reality behind them."[41] Furthermore, all human capacities are now directed toward knowledge of the "real world." They are adjusted to a *uniform*, if not clearly articulated, goal; they are converted for a *uniform*, if not clearly recognized, purpose. They become more *similar* to one another, which means that humans *degenerate* through their language. They degenerate at the very moment at which they discover an "autonomous I," become free and advance to a more "spiritualized" or "rational" concept of God and nature. "Spiritual" events, which were previously dealt with *and experienced* in analogy to events in the physical world, become ever more "subjective"; they even disappear "inward" into an active "soul" of whose laws little can be said, thereby losing their previous level of articulation and reliability as indicators of events in the outer world. The new human being has very vague new experiences. The imagist language does not disappear; it is taken over by poetry (separation of poetry and philosophy), which at first remains interested in a depiction of truth until a less stringent view of the relation between language and reality assigns fiction to it as its proper task, thereby separating poetry and philosophy/science forever. These are some of the developments accompanying the rise of "rational thought" and of Western philosophy of nature and determining the latter's most prominent features. These developments are influenced and in part caused by *historical circumstances* that at first glance have very little to do with philosophy of nature. The following examples are worth mentioning.

4.2. Historical Factors for the Emergence of Philosophy

[19] Homer's epics are embedded in older material that distinguishes itself from them in both form and content and continues to have a strong effect on Hesiod, Anaximander, and Greek tragedy. Compared to this older material (legends, myths, religious rites) the epics, despite their luminousness and vitality, are considerably cooler, more abstract, and more easily accessible for a purely intellectual experience (in the contemporary sense). As we have seen,[42] events and situations are composed of event "atoms" and situation "atoms" and articulated by analogy with familiar processes and simple objects of unanimated nature. This helps focus our view, simplifies recognition,

---

[41] This concept may have been based on the dubitable connection between *symptom* and *disease* in medicine; see Alcmaeon (DK 24A1 and 24B1) and Webster (1957: 36).

[42] See above, section [15ff.] [*chapter 3*]. See also Webster (1958: 292).

and stabilizes concept formation while rendering it "objective" in the sense that everybody in the same kind of situation is now able to *experience* the same clear impression and to *produce* the same clear description; the requirements of a *technical jargon* are already partly met, even though instead of clearly defined *abstract ideas* its elements are *concepts* distinctly *manifested by the senses*.[43] We also saw how this feature is due in part to the customs at oriental courts and in part to the requirements of poetry in the oral tradition. Thus, *one* reason for the move away from myth and toward the development of a "scientific style" (one that for the time being still merely catalogs, not essentially summarizes) lies in ceremonial rites and in certain stylistic elements of contemporary poetry.

Both occurrences on Earth and events in Heaven (whether it be starry Heaven or the Heaven of the gods) lack all emotional elements. There is no star *cult*, and religion is characterized by a cool and tolerant attitude instead of ecstatic devotion and exclusiveness; the relation to the gods is unstrained and occasionally quite brazen. Nonetheless there is a strict separation between gods and humans. Relations between the living and the dead are also more relaxed than previously; atrocities, wars, barbaric customs, human sacrifices, and narrow tribal ideologies are rare finds. Mores lack the kind of absoluteness, of moral exclusiveness, that was familiar in earlier times and later became subject to attempts at reintroduction by intellectual means after the enlightenment of the fifth century had subsided. Undoubtedly these traces are in part the result of *conscious efforts* by the *poets* themselves, who strive to convey to later generations the heroism rather than barbarism of earlier eras.[44] To a much greater degree, however, they are the result of *natural reactions to unanticipated events*. The migrations subsequent to the Dorian invasion tear apart the close connection between the tribe, the graves of the dead, the spirits bound to these graves, and the local gods bound to certain localities. Moreover, they enforce a certain general view both of human life and of the powers surrounding it.[45] Thus, *this* step into the realm of the "philosophical" does not take place out of curiosity, nor is it the result of a *conscious critique* of previous forms of life – as

---

[43] Later writers took advantage of this, such as the authors of the *Corpus Hippocraticum*, who had recourse to the Homeric descriptions of wounds, as well as Aristotle, who used observations of animals in both the *Iliad* and the *Odyssey* in his *Historia animalium*. Homer's descriptions of wounds and animals were discussed in Otto Körner (1929; 1930).

[44] Though the latter still shows up occasionally; *Iliad* XVIII, 177, 534–58.

[45] Gilbert Murray has given us an impressive account of this development in the third chapter of his book *The Rise of the Greek Epic* (Murray 1934).

if saying, "Now that we are migrating from one country to another, one island to another, the local gods can no longer accompany us. Thus we need to create more abstract and also more powerful divine helpers." Rather, it is caused by a natural development of the divine concept, which is induced and guided by special historical circumstances. Admittedly, we know very little about the *laws* guiding such development. For, after all, as an alternative reaction the gods could have been *eliminated altogether*. And yet a *conscious* definition based on a meeting of the tribal council remains very unlikely. The desire to retain old customs as far as possible in new situations, even in situations of war and tribal disintegration – which is an essentially *conservative* desire – expands the concepts without utilizing any intermediary conscious thoughts. Thus, war and confusion are *one* additional cause of the emergence of more abstract concepts in Greece.

War and internal tensions within contemporary small communities promoted the development of a more abstract way of thinking in other respects as well. In the ninth century warfare was a matter for aristocrats, who were the only ones with the necessary equipment and training. Their subordinates followed in a chaotic crowd, throwing rocks and encouraging their leaders with acclamations. The eighth century witnessed a rise in general wealth (trade; easy access to metals; new techniques in metal processing; import from the east), and hence foot soldiers were able to afford far better equipment. Political happenings in Corinth (Cypselus' "revolution") and Sparta (the offsetting of traditional oppositions between more powerful and less powerful groups through legislation) dissolved the complex ties of individuals to their ancestors, masters, and tribes. The purpose of Sparta's reform certainly was a return to an earlier situation: the predominance that had been obtained by certain groups over other groups in the course of history was to be abolished, and the earlier "equality" was to be reinstated for the wellbeing of the tribe as a whole. Yet this inherently conservative measure had revolutionary consequences. An equality required by law is different from an equality expressed in actually existing historical and social relations. And the possibility of turning such an abstract equality into an actual historical force, the very success of the Spartan undertaking, requires additional interventions in tribal life. New relations between people and new functions for the newly created groups are devised semi-instinctively, semiconsciously, and often by almost blindly giving in to the pressure of the circumstances. The hoplite soldiers were created; these no longer automatically followed the leader but participated in wars according to new strategic rules. And this also marks a gradual transition to the concept of commoners, who are defined not by their *special* relations to *special* operational units within the tribe, but by the laws of a novel abstract unit, the *state*.

Even a conscious change in the manner of presentation does not have to be based on a rational critique of the suitability of said manner of presentation for reporting objective matters of fact. Local historians of the sixth and fifth centuries, somewhat summarily classified as "logographs," wrote a style that was "clear, ordinary, pure, concise, and suited to the events and exhibiting no artificial frills" (Dionysius of Halicarnassus, *Thucydides* V, 331). We might think that this could be the result of a conscious effort to become independent of the style, and thus of the cosmology and way of thinking, of myth and legend. However, Dionysius' remarks point in a different direction: "They all had the same goal: to make generally known the traditions of the past as they found them preserved in local monuments and religious and secular records in the various tribal and urban centers, without adding to or subtracting from them." The form of such records deviates in some respects from the form of myth (though not in Homer's specific style). For those records are *lists* in which one king, god, or event follows the other. The lists contain few "poetic" elements. But why was this form *retained*? Because glorification of the Greek past, which was more characteristic of a "mythical" form of representation, was dangerous at the time of the Persian dominion in Ionia. "But simply to tell the truth, to record events such as were described in the annals of their cities, could not possibly be considered dangerous or subversive" (Pearson 1939: 16). A third reason for the development of a "strictly scientific" style, and thus of an "objective," if not yet substance-based, worldview, was *anxiety*. Furthermore, there were few myths in Ionia whose form could have been copied (Nilsson 1932: 14). Thus, an historical coincidence came to the anxiety's aid.

Multiple *methods of critique* were used to illuminate, examine, and reshape the traditional material, and they are of use in adding valuable elements to our own range of critical tools. There is, for example, the method of absurdity. It is used in the critique of the gods and takes on the shape of somewhat haphazard and very frivolous stories, later collected by Aristides of Miletus.[46] Familiar occurrences, which as natural components of our environment usually escape our attention, are *exaggerated*; they are lifted out of the flood of everyday impressions and enlarged, and thus their dubiousness is brought

---

[46] Example: *Iliad* (bk. XIV), Hera's seduction of Zeus.

to our conscious attention. Aristophanes later became a master of using this method against philosophers of nature, politicians, and rationalist city planners. The method has the advantage of preserving the audience's freedom of judgment. The audience is not shackled in conceptual chains and relentlessly dragged in a certain direction. Instead, they are merely *woken up* from their daily routine and *challenged* to take a stance, which they can then accomplish in one way or another on the basis of their *own* standards. Even Xenophanes B 28 ("[T]hat [Earth] below goes on without limit") may be an ironic critique of the view that the problem of how the world came into being can be solved by assuming that it has no temporal limits. If *this* is supposed to be a solution then the limitless downward expansion of the Earth also qualifies as a solution to the question of how it can remain motionless in space.[47] Hesiod introduces his view simply as the *truth* without even a trace of humor. The passage in which this occurs (*Theogony*, 26f.) contains a literal quote from the *Odyssey* (XIX, 203): "We can tell many a feigned tale to look like truth," as if suggesting that Homer's ideas should be put aside and replaced by superior ones. The method of critique displayed here is dogmatic assertion of a thesis. *Hecataeus*, who approaches tradition with deliberate criticism, formulates his critique in terms of a comparison of what is *credible* in his time and for people of his kind: each legend has an historical core surrounded by unlikely exaggerations and misrepresentations depending on the effect this core has on witnesses or audiences. The misrepresentations can be easily identified by their lack of credibility.

> Now according to this principle, whenever the attention an event enjoys is due to its extraordinary nature, it will be adjusted to the ordinary. If someone has achieved some fame based on the number of his sons, we will assume that this number is not greater than 20, and certainly not 50. Thus the universality applied here as a criterion is that of the ordinary, the common, the trivial.
>
> (Fritz 1967: 75)

This is the first formulation of a naïve naturalistic interpretation of older traditions. It is surprising that Hecataeus of all people focuses on the ordinary and trivial. His life coincided with the high point of

---

[47] Olof Gigon offers a similar interpretation: "Quite a few of his poems would have been of an Archilochean color" (Gigon 1968: 169). In other words, they are *mock poems* that strive to expose certain ideas by exaggeration or application in an unexpected context. Brecht offers an excellent psychological concept analysis of the method of "alienation" [Verfremdung] by exaggeration.

the era of discoveries. In the seventh century a strong east wind made the Samian Colaeus drift westward into the Atlantic ocean on his journey to Egypt, until he eventually reached Tartessos (Herodotus, *Histories* IV, 152). This was followed by the circumnavigation of the African continent by order of the Egyptian king Necho around the turn of the sixth century. After that the Persian Sataspea, sailing southward through the Straits of Gibraltar, reported sightings of small people clad in palm leaves (Pygmies – thus, he had reached the Cameroonian Highlands). Yet he decided to turn around even before he had completed the continent's circumnavigation. Euthymenes traveled south in the Atlantic Ocean and got to observe the Etesians making a river rise which he thought was the Nile. Around 519 BC Scylax of Caryanda explored the East under orders from Darius. All of these expeditions expanded the horizon and brought back reports of something extraordinary.[48] In his political recommendations at the time of the Ionian Revolt (499 BC) Hecataeus described the power of the Persians, the nations that were at the emperor's command, and advised the Ionians to build a fleet. After the catastrophe he opposed their emigrating to Sardinia or Thrace, as Aristagoras of Miletus, the leader of the revolt, had suggested, and recommended that they instead secure the island of Leros and recapture Miletus from there. Such a person should have exhibited a better understanding of the extraordinary – assuming he did not outright doubt the past's ability to produce something extraordinary – e.g., based on a *theory of progress* that would establish the present time as superior to the ancestors.[49] After all, even "modern" scientists cannot imagine that Stone Age populations might have possessed any more than the most basic knowledge of their environment.

The tragedians of Athens, by contrast, revealed the problematic character of the legends in a very different way. They did not alter the legends (except for intensifying them somewhat); nor did they "critique" them; no one commented directly on the ideology they contained. Yet they did show that the world of legends can produce situations that lead to disaster, no matter what decisions humans make. Such situations testify to the dubiousness not just of *certain actions* but of an entire interconnected *system of conduct*. The Orestes of the *Choephori* lives in a world where "there are no independent courts, nor anything at all above or independent of the monarch, so

---

[48] On the expeditions see Henning (1936) and Fritz (1967: chs. II and III).
[49] His contemporary Xenophanes endorsed such a theory of progress; see Edelstein (1967: ch. I).

that the entire order of the state rests on the monarch's legitimacy" (Fritz 1962: 123). Aegisthus usurped the throne, and only the legitimate heir could liberate the country from the murderer's dominion. Orestes continued to live in a world where the fate of humans in the afterworld is identical with our recollection. Thus Agamemnon, who is cravenly murdered, does not receive the honors that are rightfully his. He was deprived of the option to take revenge, since his corpse was subjected to "maschalismos": his extremities were removed and tied around his head.* The killing of murderers was necessary for social reasons as well as reasons of honor and justice, and was a requirement for entering a blissful afterworld. Apollo emphasizes its *objective* necessity when he orders Orestes through an oracle to avenge the murder of his father by killing his murderers in the same way. Yet among those murderers is Clytemnestra, Orestes' mother. The blood-revenge command here stands in direct and inevitable opposition to the prohibition on murdering one's own parents. However Orestes decides, he will be guilty of committing a breach of commands; there is no escape. Aeschylus treats both commands as equally strong, which sets the stage for an inevitable conflict.

In an earlier version of the *Oresteia*, which Wilamowitz-Moellendorf believed could be extracted from suggestions in texts by Simonides, Stesichoros, and Pindar, the blood-revenge command is stronger than the command of love toward one's parents, and thus the conflict does not arise.[50] Aeschylus' altered version of the situation most certainly has historical reasons: if the value of a human being lies in an inner quality rather than in the manner in which he is integrated with his environment, then his socially determined behavior shifts in favor of a more uniform assessment of individuals, thus again reinforcing the self-experience of individuals. We can assume further that Aeschylus, in the interest of maximum conflict potential (as well as maximum discontentedness with the existing social structure), staged this contemporary trend in a slightly skewed manner. Hence he could contribute to an accelerated development toward heightened self-consciousness (once the audience identifies with Clytemnestra the game has been already half won). In this way poets deliberately

---

* [*Translator's note: The Oxford Classical Dictionary defines "maschalismos" as "the practice, mentioned in tragedy (Aesch. Cho., 439), of cutting off the extremities of a murder victim and placing them under the corpse's armpits (maschalai)." So the word literally means "armpitting"; see http://www.perseus.tufts.edu/hopper/text?doc=Aesch.+Lib.+439&fromdoc=Perseus%3Atext%3A1999.01.0008. Thanks to Fiona Sewell for pointing this out.*]

[50] See Wilamowitz-Moellendorf (1896: 21ff.; 1914: 190ff).

102

contribute to building a new experience of the world, which in turn has consequences for both epistemology and our concepts of nature.

Note that this method of critique is not self-critical. Still, it encourages the critique of existing forms of life. The material the poet presents *clarifies* certain aspects of the audience's ideology, and does so in a way that makes it impossible for the audience just to rest content with what has been clarified. The audience is not relentlessly dragged in a certain direction by the thread of conceptual developments; it is not turned into a slave of "reason." Rather, the audience is challenged to take a stance on the basis of *criteria that they must select for themselves*. Thus, their freedom is incomparably greater than that of participants in a "rational discussion." Note also that even though the flow presented by a tragedy is considerably simplified compared to "real life," it does not lose its figurative nature, its effect at the level of emotions, instinct, fear, and affection. *Rather than purified thought schematisms, it is real tensions that determine the decisions* and disclose the limits of a purely conceptual procedure. For why should a compulsive logical development have greater force than the soft coercion of sympathy? On the other hand, it is not possible to extract from the procedure an abstract scheme to be applied in other areas: a sequence of events (the killing of Agamemnon by Orestes' mother) is embedded in a well-structured flow (the social background), and we observe the problems that arise from interactions between the two. The sequence of events is selected such that *all* possible solutions lead to problems. *Hence*, the social background needs to be changed. The logical scheme is this: if A then either B or not-B. Both B and not-B have undesired consequences. Therefore, not-A. Zeno later uses the same scheme in his arguments against wholes that are constituted solely by their parts.

Finally, a description of the historical background should not leave out the philosophical, scientific, and religious ideas of the *Orient*.[51] The rise of astronomy in Greece, the use of numbers in descriptions of the universe, and the use of particular numbers in descriptions of particular sequences of events all would have been unthinkable without knowledge of the Babylonians' sophisticated astronomy. The rise of the notion of consciousness and of an autonomous ego both in philos-

---

[51] An excellent example of an older discussion of oriental impacts on Greek thought is Eisler (1910). Ch. VI of Waerden (1966) contains a first introduction to the impact of oriental doctrines of the soul on Greek thought and especially on Greek astrology, with a large body of references. See also the writings in Cumont (1960). Schwabl (1958) contains a comprehensive report on Greek cosmogonies, creation myths, and their relation to oriental thought.

ophy and in self-experience was encouraged by oriental doctrines of the soul. Thus, Hesiod's and the Ionians' early cosmogonies are based on oriental ideas both structurally (succession myths, alteration as a method of generating new entities, grand catalogs, eternal nature of primary principles) and with regard to content (separation of Heaven and Earth, chaos, abyss, darkness, and moistness at the beginning of the world; self-impregnation, castration, war of the gods). Many of these ideas can be traced back beyond the period of interaction to historically obscure yet intellectually very lucid prehistoric times and may have been part of the late Paleolithic body of wisdom. Thus, the special form of life that we now proudly call "rational thought" – without having any idea of what it consists in and what its advantages are – developed surprisingly quickly, nourished by ancient traditions, contemporary factors, random effects of more recent historical events, the aftereffects of Homer's rationalism, a desire for universality, Ionian no-nonsense directness, a critical turn against legends, and new experiences of the self and the world, affecting all of their causes retroactively and uniting with them in manifold combinations. Herein also lies the beginning of our philosophical approach to nature as well as the beginning of science. The development can be reconstructed roughly as follows.

4.3. Predecessors in Hesiod's and Oriental Cosmogonies

[20] The starting point is *narratives* of the *origins* of certain basic conditions of the world such as the Earth, Heaven, Ocean, followed by gods and various demons (the gods come later since there must first be a place for them), as well as of the *later developments* of these principles. In the Orient the basic principles have both a personal and a material side, which together explain their characteristic perceivable behavior. In addition, the personal components also enable us to obtain a direct understanding of obvious facts. The principles originate either by birth – in which case some particular power, such as Eros, is often the mediator – or by mere isolation, "without sweet union of love," as Hesiod puts it (Hesiod, *Theogony* 132). In the latter case we often encounter a similarity in content between the isolating principle and the isolated one: the process of isolation is subject to a principle of conservation. What is conserved is either a quality (which later, after the removal of the figurative component, leads to a purely conceptual deduction) or a number: a system of rules assigns each word denoting a basic principle a certain number, its *psephos*

(a predecessor of the Gödel number for sequences of symbols). Isolation processes or general changes proceed such that both the starting point and the end point of the process have the same *psephos*.

The description of changes is accompanied by a description of the gradual development of a cosmos, that is, of a well-defined, well-structured world stage as a setting for the history of gods and humans.[52] Now, this history is not of a kind that knows no bounds, nor is its background only vaguely indicated as in Homer; instead it takes place in a clear and plain environment. The latter, though, may in turn be embedded in an amorphous ocean of a basic principle (chaos in Hesiod, the *apeiron* in Anaximander) from which it emerges, in which it rests, and into which it may eventually dissolve. The cosmos is subject to a uniform *law*. At first this is the law of a victorious god who, after subduing opposing powers hostile to the order that he imposes, introduces the latter as the only valid order of things (Marduk in the Babylonian creation myth; Zeus in Hesiod's epic). Subsequently that law becomes independent of an individual's *dictums* and turns into a law of *right* immanent in the universe (Anaximander, Heraclitus).

In almost all oriental sources the beginning of this development is non-temporal. Though Hesiod recognizes emergence in time, it is eternity again in Pherecydes, Anaximander, and the Ionians. Personal gods disappear in the Ionian philosophy of nature, but they leave their traces; qualities continue to have a subjective and an objective side, and for a long time the cosmic order remains a legal order. The disappearance of the gods is facilitated by the minor part that they play in Homer's *physical* cosmos as well as by the Ionians' entirely non-mythical imagination and experience of the world. Today this step is generally regarded as "rational," and the gradual elimination of divine traces as a further increase in rationality. Yet this means identifying rationalism with materialism – a dubitable procedure based on a naïve naturalistic interpretation of the material. It indicates an oversight of the possibility that materialism may have *contradicted* the contemporary experience of the world, and so it may be considered "irrational" in light of an empiricist methodology. Let us also remember that the Ionians' world experience was more materialistic than that of their predecessors in Greece and in the Orient, making it *less sophisticated* than the latter. But why would choosing the less sophisticated over the more sophisticated experience have been more "rational"? (Please note that we are talking about experience, *not*

---

[52] Regarding the concept of cosmos, see the examination in appendix 1 of Kahn (1960).

about speculation!) And today we finally do recognize the animated nature of the world, the existence of creative powers even in inanimate matter, which in an uninterrupted flow continuously introduces new life forms into the world.

Not only the flora and fauna but also matter and cosmos as a whole are links in an *evolutionary* chain that affects everything. Space, time, and matter change along with the laws that are used to determine them. Creative shifts – leaps from quantity to quality, to use a more objective terminology – are standard. Each part of the world has a *history* that explains its structure. Developmental sequences connect animated and unanimated, irregular and regular things; they extend from cosmic catastrophes, to periods of rest in which enormous powers maintain equilibrium, to further catastrophes caused by a disruption of equilibrium. Modern cosmogony and cosmology (which includes the history of life and thought) concurrently repeat the general principles of the old mythical theories. Even the modern theory of catastrophes and intermittent equilibrium has a matching mythological analogy in the story of the War of the Titans and of the taming (not elimination) of the Titans by a powerful god (so that the subsequent order is a *dynamic* order, which can lose its equilibrium at any time). These facts need to be taken into consideration if we wish to accurately evaluate later Greek philosophy of nature and its ventures. These ventures result in rejection of the development scheme as conceptually unsatisfactory and its replacement by the scheme of deductions from axioms. Rigid forms, clear distinctions, and unequivocal concepts replace the manifold of dazzling powers, of which motion, transition, and productivity are essential characteristics. Only in a very indirect manner can such necrotic material serve to reconstruct the reproductive, living world. A true odyssey has to pass through our thinking, through a long chain of errors, before it can re-approach the real world and recognize traces in it that were once described so vividly by myths of creation and development. The odyssey starts with *Parmenides*.

The impact of Parmenidean thought on oriental philosophy and science can hardly be overstated. Parmenides was the thinker who replaced concrete sequences of events with invariable and purely conceptual laws, thus strictly separating experience of the world from reality, intuition from thought, knowledge from action. Mathematics owes him the turn from the intuitive to the abstract and its "rise" from a general theory of the life world to a theory of ideal entities. Science is indebted to him for its belief in eternal laws and its axiomatic method of representation, which has now come to be regarded as the universally valid basis of understanding. He was the starting point for all *problems of justification,* which are but problems of the relations between those domains that are now so radically separated from one another. What was once harmoniously connected gets separated, then subsequently all mental capacities are employed to find a way to reunite the two. Parmenides also was the starting point for all attempts to understand the intuitive world and its reciprocal effects, changes, and transitions on the basis of principles that at first glance have nothing to do with it. Thus we step away from the world, rebuild our thinking, turn a multi-use adaptation mechanism into a simple, shiny, easy-to-understand computer, and then re-approach the world with this new thinking tool. The theories that emerge as a result fall more or less naturally into two groups.[53]

The first group of theories attempts to solve the problem of motion and development by means of concepts and entities that do not have any direct counterpart in everyday experience and whose description often considerably deviates from ordinary language. Thinkers associating with this group of theories have a hard time connecting their theories with perceptual evidence, and often can do so only in a very schematic way. Examples are Democritus, Plato, and their successors in the modern sciences. The second group of theories attempts to grasp the objects of our experience in a more direct manner and to show that the concepts of ordinary language used in this undertaking are not too far removed from the Parmenidean ideal of invariability: some minor changes suffice for them to satisfy the requirements of a scientific report. The most prominent thinker associated with this group is *Aristotle,* the preeminent creator of a qualitative theory of nature. Let me now start with a more detailed account of certain phases in this development.

[21] Following the temporary neglect of genealogical schemes, Hesiod proceeded to place world events back into an historical context. He took his preferred organizing principle – the *genealogical tree* – out of the grammatical structure underlying his narrative and made it explicit, thereby linking animated and unanimated aspects and embedding both in a more comprehensive context. All these traces suggest an oriental influence. The creation starts with the emergence of Chaos, of the Earth together with "dim Tartarus," and of Eros. Chaos gives birth to Erebus and the Night, and the latter together with Erebus to Bright Air (Ether) and Day. The Earth brings forth the starry Heaven, Mountains, Meadows, Fields as well as Depth and Ocean, though the latter "without sweet union of love" (*Theogony* 132). With that the *physical universe* is complete.

---

[53] See below, section [29] [*chapter 6.1*].

The following characteristics of this narrative deserve special attention: first, the world is not described in terms of abstract principles from which the manifold of the cosmos *logically* follows; thus, the account is not "based on postulates." Instead, it assumes an undifferentiated original state from which the world as we know it gradually and in consecutive steps *developed*. Such a *narrative philosophy of nature* was widespread even prior to Hesiod. It is the very form of description of the world that is characteristic of many myths. Almost all mythological accounts of nature and society tell a story about how a particular state developed; they are not content with a mere *analysis* of the state. In that they are actually closer to our contemporary level of knowledge than the ideology of "eternal" laws of nature and "invariable" basic components (of matter, of life, of society). We have realized that the notion of an invariable basic law or an absolutely stable element (atom, gene, etc.) only has approximate validity, that the basic form of matter is motion, that this motion governs all areas of life, connecting animated and unanimated things, and that it governs the universe as a whole with all its laws. Hesiod expressed all of these ideas with unsurpassable clarity. Homer presupposed the successions described by Hesiod, but he shares the underlying worldview only indirectly via an account of *structures*. The Ionian philosophy of nature, especially in Anaximander, uses Hesiod as a mediator to have recourse to older forms. (It makes use of the same initial states of development, which correspond precisely to the four elements.)

The second remarkable characteristic is the *materialism* used for the part of world history discussed above. As Cornford puts it in a rationalist exaggeration, Chaos, Earth, the starry Heaven, the Ether,

> are not here supernatural persons with mythical biographies and adventures. Even when Earth is said to "give birth" to the mountains and seas, Hesiod himself tells us that this is conscious metaphor: a "birth" can only follow upon a marriage, but here it occurs "without love or marriage", *ater philotētos ephimerou* [*Theogony*] (132). The metaphor means no more than that this cosmogony is of the evolutionary type. [. . .] [T]he personal gods come later.
>
> (Cornford 1950: 96f.)[54]

---

[54] But see Cornford's correction further on in the text: "[C]osmogony [. . .] is not a myth [. . .] It has advanced so far along the road of rationalisation that only a very thin partition divides it from those Greek systems which historians still innocently treat as purely rational constructions" (Cornford 1950: 100; see also 1952: 198). [*Translator's note: Feyerabend's German translation of this has the following slightly changed ending: "[. . .] that only a very thin partition divides it from the early Ionic systems."*]

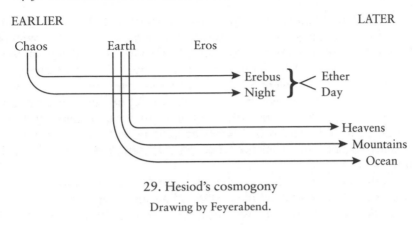

29. Hesiod's cosmogony

Drawing by Feyerabend.

This becomes obvious especially in the *reification* of the concepts. In the sixth and fifth centuries BC the word "chaos" roughly means hollow, disconnectedness, gaping abyss. Thus, where there is a "chaos" there also have to be certain other things in which it creates a hollow, which it disconnects or whose gaping abyss it is. This does not match the course of events as described in Hesiod. In his account Chaos *came first*, followed by the "wide-breasted Earth," the Ether, the starry Heaven, and so forth. Thus, "Chaos" here does not have the relational meaning indicated in the word "abyss"; rather, it *has to be of a more substantial nature.* This substance lasts even after the world and the gods have been created and surrounds the stage of history as one "great gulf" (*Theogony* 740) the bottom of which is beyond anyone's reach. Material substances have a general tendency toward self-preservation. And indeed, those first transitions follow a "principle of conservation of qualities." Chaos is *similar* to the Darkness and the Night, which it produces, while the Heaven, Mountains, and Ocean are similar to the Earth.[55]

Despite this concurrence, the "materialism" displayed in *Theogony* 116–32 is fundamentally different from later materialistic theories. (Cornford overlooked this fact.) The materialism in *Theogony* is only one *side* of the primordial powers; it occasionally guides the creation process, but it is not their *essence.* Right after the birth of the Ocean "without sweet union of love," the Heaven and Earth unite and, now

---

[55] This similarity convinced Schwabl to regard Erebos and Night "as properly belonging to the 'concept' of [Chaos]" (Schwabl 1958: 1440).

as the persons Uranus and Gaia, proceed to give birth to a series of individuals including Rhea, Themis, and Mnemosyne, "gold-crowned Phoebe and lovely Tethys," as well as "Cronos the wily, youngest and most terrible of her children" (*Theogony* 135f.). The diversity of the primordial powers is reflected in the diversity of the words used to describe them, which are not clearly defined but rather point to one meaning at one time and to another meaning at another, thus avoiding the obstruction of further development. It is not wrong to consider Hesiod's materialism a *dialectical materialism* (in theological clothing).

Finally, the introduction to the *Theogony* contains important and influential *epistemological* ideas. Hesiod actually presents his report dogmatically as the *truth*: "We know how to speak many false things as though they were true; but we know, when we will, to utter true things" (*Theogony* 27f.). It is notable here that the narrators remove themselves from the context of the narrative and confront it. They maintain their role as free creators – who can either create fiction or tell the truth and who *choose* to employ the power of their words in the service of truth – rather than as merely correct reporters. Truth is no longer a necessary quality of the narrative but a consequence of a contingent event, namely a decision. Just as in Homer, the Muses are the narrators here – Hesiod listens to their words and repeats their doctrine – but the Muses introduce the problem of fiction in a new manner: each part of their report could be false, if they chose it to be so. Yet they did not make this choice, and hence each part is true.[56]

Thus Hesiod's report, even where it is true, is no longer needed, and the desire for assessment becomes understandable. And so it did not take long for Xenophanes to challenge the entire ideology behind the genre of epics. Thus the critique was prepared in mythological clothing. Yet for the time being another element of the epic form of story-telling was still preserved: as in Homer, knowledge is presented as a *list* of truths and illusion as a *list* of falsities. Thus a fictional tale from the Muses would not be a *general illusion* but rather the narration of a *number of false things* – "*idmen pseudea polla legein*" (*Theogony* 27). This characteristic, which is connected to the persistence of the formula structure, disappeared completely only due to Parmenides.

The powers created upon completion of the first part by a union of the principles Uranus and Gaia, now displaying their *personal side*,

---

[56] In the *Works and Days* Hesiod also narrates things that are restricted neither by personal or communal experience, nor by what is accessible to the gods, hence things that are entirely arbitrary; this is how some authors interpret the word "athesphaton" (*Erga* 662).

110

have an important function. Though this brings us back to "that world of mythical representation which the rationalised cosmology had left so far behind" (Cornford 1950: 100), the "return" introduces aspects that would be neglected for a long time afterwards until their gradual revival in late nineteenth-century science and then again just recently in philosophy of science. The *world* as we know it today with its laws and, at least in the domain of nature, its regular processes is a product of a *dynamic interchange* in which origination processes are sometimes suppressed and at other times take on the form of disasters. The *concepts* of a prolific science are not as well defined as would be desirable from a "rational" point of view, *and they ought not to be* if they are to live up to their task of a stepwise adjustment to complex circumstances that are *affected* by many familiar and unfamiliar factors. The story of the gods' wars meets both requirements. Just as in the later geometrizing of nature, the most important terms here are *technical terms*[57] whose logic is used to uniquely describe certain structures conjectured to exist in the world, yet in a manner that not only avoids removing the possibility of later developments but makes them accessible through the nature of the very terms used (here, in the area of change, geometry has proven to be hopelessly incomplete). From this perspective the sons and daughters of Uranus and Gaia are forward-pushing trends (cf. the etymology of "Titans": the dragging, stretching ones) that are temporarily restrained by their father. The castration of Uranus, which is obviously nothing but a special version of the widely spread myth of the separation of Heaven and Earth (Staudacher 1942: part II), breaks this obstructive force and makes room for new developments (sun, moon, and stars are created just at that moment – *Theogony* 371ff.), even though the laws of the world are not yet finally determined and secured. That happens only in the course of the War of the Titans, which occupies the entire world stage (*Theogony* 695ff.) and culminates in Zeus' victory. Zeus introduces a new world order. He establishes the laws of the world as we know them today. Zeus' later coming into power has important consequences: "Only because this god Zeus, as the highest god, is not at the same time the oldest god but rather the leader of the younger generation this theogony could become the origin of cosmogony and with that of the later science of nature [with its "eternal laws"]. If Zeus had been the oldest god [. . .] then the creation myth could have added new gods at any time" (Krafft 1971: 75), and the new order would never have been secured. It is this order that would later be

---

[57] See section [11] [*chapter 2.1*], second approach, and section [12] [*chapter 2.1*].

separated from its mythological background to become the basic element of Ionian philosophy of nature.

The idea of a separation of Heaven and Earth also experiences a later revival in the form of the separation of the elements that constitute the structure of the world – in particular, the elements located *between* the fiery Heaven and the firm Earth: water, clouds, mist, air. Even Aristotle regarded heat, cold, moisture, and dryness as primary qualities constituting the four primary bodies, the elements: "All other differences, such as heaviness and lightness, density and rarity, roughness and smoothness are secondary, 'for it seems clear that these [the four primary qualities] are the causes of life and death, sleeping and waking, maturity and old age, health and disease; while no similar influence belongs to roughness, smoothness and the rest.' A long history, stretching back into the mythical epoch, lies behind the statement that this reason for their primacy 'seems clear'" (Cornford 1952: 197). And finally, Hesiod anticipates later notions both of a well-arranged stage of world events, a cosmos, and of certain elements of this stage: a disk-shaped Earth with a bordering Ocean flowing all around her, covered on every side by a Heaven "equal to herself" (*Theogony* 126f.) and with a falling distance of nine days above her (*Theogony* 722f.), and this Heaven continues on below into the equally deep Tartarus, so that we have a disc surrounded by two hemispheres. The entire structure is embedded in the infinite Chaos. Thus, these are the elements that were directly accessible to the first philosopher of nature, Anaximander of Miletus. Let us see what use he made of these elements!

— 5 —

PHILOSOPHY OF NATURE
THROUGH PARMENIDES

[22] It is customary to start an account of the history of philosophy (of nature) with *Thales'* theory according to which the world and everything in it is either made up of or generated from water. Any such account would rest on a quote by Aristotle, who in his *Metaphysics* 983b18ff. refers to Thales as "the founder of this type of philosophy," where "this type" includes comprehensive *principles*, all of which are *material*. A more detailed look at the tradition, however, leads us to conclude that Thales still compiled catalogs of the most astonishing events such as flooding of the Nile, solar eclipses, earthquakes, and magnetic properties, which also contained explanations for these facts. These explanations were not coordinated with one another; their sole purpose was to integrate each event, taken *in isolation*, into a larger context, thereby making it comprehensible. Thus Thales appears to have interpreted the origination of land in Lower Egypt as a gradual emergence of the Earth disk from the surrounding water, until it finally floats in the water like a ship. Accordingly, earthquakes are nothing but risings and fallings of that ship, and the springs that burst open during earthquakes are leaks (Aristotle, *Heavens* 294 a28ff. – note that the problem is not the stability of the Earth as a whole, but how to explain certain events occurring on its surface). The broader context is almost always of a mechanical nature, though it is possible that for Thales mechanical effects may ultimately be due to souls or gods: "Thales said the mind in the universe is god, and that all is endowed with soul and full of spirits; and its divine moving power pervades the elementary water" (Aristotle, *Soul* 41 la7ff.). Yet the following assessment is almost certainly true:

> Thales himself did not belong to the old natural philosophers because he
> pursued the question of the origins and totality of the world with special

speculative forcefulness. In this respect Hesiod and Anaximander were vastly superior. But he saw the world and described what wonders he found in it. What makes him a philosopher is the noticeable will to find an intelligible cause in each individual case, which would remove the unearthly appearance from the wondrous matter and make it familiar to us.

(Gigon 1968: 58)

The familiar are the *mechanical* things that we know from our immediate environment but that are also pervaded by divine power and therefore causally active.

5.1. Hesiod and Anaximander: Changing Worldviews

[23] *Anaximander*'s own worldview is composed of elements of Thales' list of explanations together with his own knowledge and principles borrowed from Hesiod.[58] Just as in Thales, the Earth emerges because the water above it disappears. In contrast to Thales' worldview, the water disappears because the sun makes it evaporate (Aristotle, *Meteorology* 353b6ff.), and not because the Earth rises above it. We may assume that this part of the story originated in the course of criticism of Anaximander's much-traveled fellow denizen; Anaximander appears to have been far more settled than Thales. Another contrast to Thales' worldview is the fact that explanations are only a small component not of a list but of a comprehensive *development* of the world and the events in it, striving to make comprehensible numerous wonders as well as everyday events by drawing on similar and coordinated basic principles. The starting point of this development, the *apeiron*, shares characteristics with Chaos in Hesiod; it is indefinite and unlimited, but unlike in Hesiod it is "ungenerated, immortal, and indestructible" (Aristotle, *Physics* 203b18-20). In this respect Anaximander had a forerunner in Pherecydes (DK 7B1), a contemporary of Thales, although the notion of the eternality of prime powers was widespread in the oriental region. The Babylonian creation myth, for example, assumed three prime powers: Apsu, Tiamat, and Mummu (personalized sweet water, equally personalized saltwater, and the mist above both).

In the myth the epithets "immortal and indestructible" are attached

---

[58] On Anaximander see the comprehensive account in Kahn (1960). Criticism of the sources used in this book can be found in Schwabl (1964, 1965). Dicks' diatribe (1970: 45) should be disregarded.

to the gods, and they still exhibit the epic meter as well. This may be because Anaximander regarded the gods as *symbolic* representations of natural phenomena, or it could be a continuation of the assumption that all basic principles of nature are divine principles. The symbolic view would later become very popular in Greece. It is hard to say whether Anaximander (and Pherecydes) held this view. It tended to be more popular with the "rationalists," but Thales (DK 11A22, 22a, 23), by contrast, held that divinity is *added* to the phenomena of nature. And Newton's example shows that rationalists were very well inclined to assume a nature governed by divine laws and to defend this view with empirical-rational arguments. Thus it wouldn't be wrong to assume that for Anaximander, too, divinity was the cause of motion and creative power, especially in light of the quote that some authors consider as a literal repetition and according to which the *apeiron* is "encompassing and directing all things" (DK 12A15): the *apeiron*, despite being indefinite and unlimited just like Hesiod's Chaos, is full of possibilities and powers and responsible for the orderly progression of events; it is "indefinite" only in comparison with the qualities and things that emerge from it.

In the only direct quotation, the orderly progression of events is represented according to the image of a community governed by the rule of law: "Whence things have their origin, thence also their destruction happens, as is the order of things; for they execute the sentence upon one another – the condemnation for the crime – in conformity with the ordinance of time" (DK 12B1). This idea, too, was anticipated in Hesiod. Gods who intend to obstruct the natural progression of world events are eliminated. That is their punishment. Only Zeus' dominion, resting on an acknowledgment of the causal powers in the natural world, is permanent. What the gods are punished for is their improper behavior, *not* their existence. Likewise we can assume that in Anaximander it is not the existence of things, as later interpreters influenced by Christianity have assumed, but rather their improper behavior that has to be restrained. As the world steers toward a form of organization, encompassing it unperturbed by death or destruction, the *apeiron* clearly enough reveals those divine characteristics that are needed to understand the endless and yet regular changes in the world. The characteristics are *abstract*; they're not associated with any particular *person* – which would furthermore exhibit randomness, occasional irrationality, and bullying behavior. They combine the abstract regularities of Hesiod's world, which – being superior to the gods – restrict all excess, with certain characteristics of the gods thus constrained. This paves the way for

115

Xenophanes' god, who is entirely separated from physical matter, keeping it under control only from the outside (that is, from beyond matter but not necessarily from beyond the world itself).

It is not certain whether along with the *apeiron* Anaximander assumed additional, independent principles or whether he thought that everything was generated from the *apeiron* by a single process. Some authors[59] conjecture that for Anaximander the Earth, the sky, and the infinite motion of the *apeiron* – just like the Earth, the sky, and Eros in Hesiod – exist independently of the *apeiron* (Chaos), while others hold that the Earth gradually emerges at the center of the prime substance.[60] The latter view is inconclusively supported by Aristotle (*Meteorology* 353b7ff.), who distinguishes between those who let everything emerge around and for the sake of the Earth and the more "secular philosophers," who assign the Earth a beginning in time but describe the actual emergence as an evaporation of the moisture "in the Earth's vicinity." Either way, a *germ* emerges within the *apeiron* that is capable of generating heat and cold or, to use an earlier wording, darkness and light, night and illumination. The germ generated a sphere of fire that "grew around the vapor that surrounds the earth, like bark around a tree" (DK 12A10). Next comes the formation of circular tubes made of air and filled with fire that "surround the earth like wheels with hollow rims" (DK 12A22) and whose open ends appear to us as the sun, moon, and stars. The solar wheel is the largest, the second largest is the moon, and the wheels or tubes of the fixed stars rank third (DK 12A11, 22). Both the solar and the lunar wheels "are atilt" (DK 12A22), by which is probably meant that they form an angle with the equator. The degree of slant may have been determined by means of a gnomon, something that Anaximander was familiar with (DK 12A2). The thickness of the wheels was equal to the diameter of the flat part of the Earth's surface, and the diameter of the wheels is 19 or 27 times this Earth's diameter. Rain, wind, clouds, thunder, and lightning are generated by the interactions between the vapor surrounding the Earth, the sun affecting the moisture, and the clouds, which burst when filled with vapor from within. Animals emerge from moisture and humans from fish, since they cannot care for themselves in the first few months of their lives. In this way the narrative combines primordial events with concrete

---

[59] For example Gigon: "Earth and sky do not get generated from the indefinite, but are there all along" (Gigon 1968: 78, see also 84).

[60] Kahn (1960: 87), with reference to DK 12A10, 33f.: "the Earth is mentioned only by anticipation, because the dry land has not yet arisen out of the central vaporous mass"; see also Kirk and Raven (1957: 133).

and current events in the world, and we obtain a clearer outline of the stage of world history than was customary in earlier narratives. Reports confirm that Anaximander not only described this outline but also used models to illustrate it, namely a globe (DK 12A1, 16) and a map of the Earth (DK 12A6).

The Earth's design itself combines the hitherto-common disk shape with the results of a radically novel theory: it stays in its location because it is the same distance from every single point in the sky. This theory presupposes that the Earth has "another side" (DK 12A17, 7), and it is only natural to locate this "other side" at a certain distance from the side on which "we stand." The distance is a third of the diameter. Thus we have a cylindrical Earth not unlike a tree trunk whose surface Anaximander appears to have described in detail.

A preliminary comparison with Hesiod reveals the following: the form of presentation is still that of an historical narrative. The elements of the narrative are linked by "principles of conservation" that have more binding force than Hesiod's and are supplemented by a control-and-equilibrium principle, which Hesiod had only vaguely indicated. The primordial power is eternal and indefinite compared to the qualities of darkness and light that it generates; and yet it is not entirely without properties. It is divine without having any personal nature.

The primordial power generates entities that have tradition-ally been called coldness and heat, though these are certainly later designations of those entities, which show up in Homer as *aēr* and *aither*. Both are originally neither locations nor substances but rather circumstances or dynamic powers. *Aither* is the brightness of the clear sky, be it that of day or night. *Aēr* is darkness, vapor that occasion-ally becomes so thick that it has to be cut through by force (even today we still say "you could cut the air with a knife") or could be used as a pad: Ares in *Iliad* V, 336, places his lance on *aēr* the way you would place an object on the ground. Thus, *aēr* here has nothing to do with the atmosphere that we breathe; it is not a body, though the boundaries between *aēr* and *aither* constantly shift, given rise to ever-new conditions of their visibility. *Aēr* in Hesiod, where "at dawn a fruitful mist is spread over the Earth from starry heaven upon the fields of blessed men" (*Erga* 548f.), is more substantial. The vapor rises up from rivers, is lifted high above the Earth in storms, and turns into rain, wind, or clouds in the evening (this brief passage anticipates a major part of Anaximander's meteorology as well as of later theories of the cycle of elements). This new meaning of the term *may* have been connected to observations in the poet's home country's climate as well as to his genetic way of thinking (without which the

117

observed stages would follow one another without being connected in any way) and with the increasing tendency to objectify. We may assume that Anaximander posited *this* pair of opposites rather than the already quite specific elements "heat" and "cold" of the later philosophy, and his tubes do indeed constitute darkness encompassing light. The separation of a part of this darkness in the near vicinity of Earth and its upward boundary by fire later results in the entirely new concept of *earthly atmosphere* and *air* in our meaning of the term, as an essential substance or condition. After the tree-trunk Earth this was the second chapter of a constitution of objects with the help of new ideas: the "air" is not discovered by experience but invented by speculation.

Two comments are appropriate in this context. First, we notice that these early philosophers did not tend to focus on what we would consider today as "natural," "familiar," or "non-threatening" (without quite knowing what we mean by such designations).[61] Earth floating isolated at the center of the galaxy was certainly not "familiar" to Greece in the sixth century (is Earth rotating at high speed around the sun "familiar" to us?). Nor is it suited to reduce fear, unless one believed in its stability as passionately as one believed in the power of the gods. For contemporaries a world without gods was not "natural." Matter rotating in isolation in between Heaven and Earth without the help of gods would have been a terrifying monster and certainly not "familiar": how can one rely on something that does not have an iota of reason? A smooth materialism was everything but "natural" for Anaximander's contemporaries – *and this for good reasons*, for they simply did not experience the world that way.[62] Again, Gigon's remark reveals the underlying naïve naturalism that equates the Ionians' experience of the world with ours and at the same time reduces "our experience of the world" to the very bleak laboratory experience of our civil servants of thought. Second, the archaic experience of the world is reflected in an elastic archaic terminology. As W. K. C. Guthrie puts it in volume I of his massive work *A History of Greek Philosophy*, "Let me repeat that we are not at a stage of thought when clear distinctions between different uses of the same word are possible" (Guthrie 1962: 86). This is an *advantage* compared to our situation today, where we first create concepts that are clearly distinguished from one another and then get into trouble trying to use these discrete entities to

---

[61] See the Gigon quote above (Gigon 1968: 58); section [22] [*this chapter, introductory section*].
[62] Nilsson (1940: ch. 1).

represent things in nature that are seamlessly related to other things. (As Bergson has shown, this is also one of the problems with grasping time.) The Ionians and their ancestors, such as Hesiod and the oriental creators of myths, were conceptually much better equipped. And due to its ambiguity their speech was a good deal more stimulating than the frozen jargon of our contemporaries.[63]

A third example of a purely speculative creation of visible objects is the *celestial spheres* and their relative positions. Anaximander introduced the concept of an *orbit*, which had such an important role in physics from Eudoxus through Kepler and onward to the "stationary orbits" of earlier quantum physics: unlike in later authors, the celestial bodies – including the planets (knowledge of which is not conclusively exhibited in Anaximander) – are not carried aimlessly, albeit with a certain regularity, by external forces. Rather, as Plato wrote in reference to the discoveries of contemporary astronomy, "each of them travels in a circle one and the same path, – not many paths" (Plato, *Laws* 822a). A celestial body's complex motions are combinations of its sphere's rotation and the sphere's motions *as a whole*, which anticipates the later "rescue of phenomena": here, too, a basic motion is assumed, and we "rescue" not the motion *as a whole* but the observed deviations from this basic motion.

The fixed stars are at the bottom level within the order of celestial bodies, something that has been often seen as indicative of the primitive nature of Anaximander's astronomy: did he not observe the way fixed stars were hidden by the moon? This objection assumes that scientific theories never contradict clear evidence – a fairly naïve position. Just like Anaximander, the father figures of later science – that is, Galileo, Newton, and Einstein – paid little attention to striking discrepancies between observation and theory so long as their positions were coherent, productive, and in harmony with certain not always clearly articulated metaphysical assumptions.[64] For Anaximander the tendency of fire to move upward may have been the trigger: the largest quantities of fire, that is, the sun and the moon, are located at the periphery.[65] This is just a brief summary of the main ideas in

---

[63] Niels Bohr is the only modern thinker who recognized this issue, which is why he insisted on a more elastic and ambiguous jargon in physics. See Feyerabend (1968; 1969a).

[64] One of the would-be experts who deny Anaximander's "astronomical competence" is D. R. Dicks (1966: 46), whom we have already encountered in a different context (section [6] [*chapter 1.4*]). In both contexts Dicks exhibits a lack of imagination combined with linguistic narrow-mindedness while trying to lift up what he considers science far above its allegedly entirely nonsensical predecessors. On the treatment of experimental problems in science see Feyerabend (1972a, 1972b: section 5).

[65] On this interpretation see Kahn (1960: 90).

Anaximander's philosophy of nature. This summary (which neglects other ideas, such as the notion of infinitely many worlds and periodic catastrophes) suggests the following *interpretation.*

Anaximander made use of material from Thales and Hesiod and combined this material coherently. He removed the personal gods' individual interventions, attached more importance to principles of conservation, and also paid more attention to detail than did his predecessors. He was "rational" on the basis of his materialism, his sense of order, his criticism of earlier positions; and he was "scientific" on the basis of his attention to detail and his introduction of concepts that would later play a prominent part in science. This interpretation is not unapt, but it neglects a change that places the then rapidly growing philosophy of nature in a new light. I shall explain this change by means of a seemingly trivial and redundant question.

[24] For Hesiod the Earth is a disk with the sky rising above it and surrounding it entirely; beneath it – at the same distance as the sky above – lies Tartarus. Anaximander said of the Earth, "We walk upon one of its surfaces, the other being opposite" (DK 12A11). Hesiod did not say this explicitly, and – as I should like to add – he did not need to do so. It goes without saying that a disk has two sides. So why did Anaximander write down something so obvious? Why the fuss?

Admittedly, the underworld's topography is not very clear in Hesiod's work. Indeed, *all* early texts are vague about the journey from the Earth to the underworld. The trip can be filled with interesting events (such as Ishtar's descent to Hell with its accompanying striptease), but *where* these events take place as well as *what relations exist between the locations* is hard to determine. Plato said: "The path [to the lower world] is neither simple nor single" (*Phaedo* 108a). Thus, a detailed account would certainly be desirable. Yet the detail that Anaximander added to Hesiod's account was superfluous. We know that a disk has two sides of which one is the one and the other the other without needing him to tell us that. That is so unless the region of Tartarus is *in principle* different from ours, as a room in a dream is different from a room experienced in a waking state. Now, later cosmologies often assumed a systematic difference of regions. For Plato (*Timaeus* 33a, b) the spherical world encompasses everything that exists. Beyond it there are neither physical bodies nor void, as Proclus emphasizes in his comment.[66] According to Aristotle (*Heavens* 279a18ff.) there is neither place nor void nor time outside

---

[66] Proclus, *On the Timaeus* II, 73, 89, 91; see also Cornford (1937: 57).

the heaven: "Hence whatever is there, is of such a nature as not to occupy any place, nor does time age it; nor is there any change in any of the things which lie beyond the outermost motion." As a consequence, any area in the outermost part of the heavenly circumference is *nowhere* and *in no time* relative to its *external* environment beyond the circumference (which according to the Aristotelian scheme would have to assign it a place). It is distinct from the surface of a ball rolling on a table. It does not have a "surface" in our familiar sense of the term. Does this distinctness also hold of Hesiod?

In most cases pictorial representations of the Aristotelian world tend to be inadequate, as can be seen in figure 30 (which is why later, more insightful people introduced a prohibition on pictorial representations). A better case can be made for the tale (which is independent of Aristotle) of Enoch's ascent to heaven (Charles 1896: 25ff.), where he loses his orientation in the top sphere and notices only qualitative things: luster, light, and the dreadful face of the Lord. In general, all verbal representations of the outermost part of the Aristotelian world, as long as they remain vague, are superior to pictures, with the exception of pictures showing a very different event space on the external circumference than on the inside, such as figure 31 from the 1661 *Harmonia Macrocosmica* by Andreas Cellarius.

Nor can Archytas' thought experiment (DK 47A24) of standing in the outermost corner of the universe with one hand stretched further out be used against the later Aristotelian or the earlier mythological view, as we shall see. According to Aristotle (*Heavens* 278b31ff.), Archytas, whose body consists of the four elements, would be capable of reaching the periphery neither on his own nor with external assistance – it would be contrary to the laws of physics. And if he did succeed in undertaking the trip, it would not provide him with any indications of location and would therefore be in vain. We can already see here that the geometric arguments presented by later thinkers are not necessarily able to refute the earlier views.

Homer's stories about the underworld in book 11 of the *Odyssey*, as well as other related stories, indicate that the descent to Hades requires special preparation and leads into a special environment in which special and strange events take place. The descriptions are precise, but in a way that makes it difficult for us to form an image or picture of the spatial arrangement. The entrance to Hades lies in the far West and requires a long voyage by sea.[67]

---

[67] Eastern underworld legends also locate the entrance to the underworld far in the West – see Heidel's commentary on the Epic of Gilgamesh (Heidel 1949: 171).

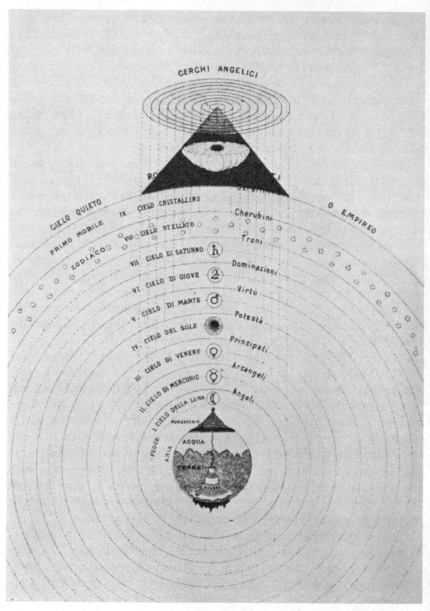

30. Misleading representation of the Aristotelian world

From: Sarton, George (1947): *Introduction to the History of Science III, I: Science and Learning in the Fourteenth Century*, Washington, p. 486 (based on Dante's model of the universe).

31. Adequate representation of the Aristotelian world

From: Kearney, Hugh and Kurt Neff (1971): *Und es entstand ein neues Weltbild: Die wissenschaftliche Revolution vor einem halben Jahrtausend*, Munich, pp. 10f. (from Andreas Cellarius, *Harmonia Macrocosmica*).

The location is:

> wrapped in mist and cloud. Never does the bright sun look down on them with his rays either when he mounts the starry heaven or when he turns again to earth from heaven.
>
> (*Odyssey* XI, 15–18)

Normal actions have strange effects:

> So she spoke, and I pondered in heart, and was fain to clasp the spirit of my dead mother. Thrice I sprang towards her, and my heart bade me clasp her, and thrice she flitted from my arms like a shadow or a dream, and pain grew ever sharper at my heart.
>
> (*Odyssey* XI, 204–8)

123

And only death can make this location accessible in a *regular* manner, since

> [h]ard it is for those that live to behold these realms, for between are great rivers and dread streams; Oceanus first, which one may in no wise cross on foot, but only if one has a well-built ship.
>
> (*Odyssey* XI, 156–9)

In his description of Tartarus Hesiod repeats the above-quoted phrase about the sun (*Odyssey* XI, 15) word for word and applies it to the dwellings of the "children of dark night, sleep and death, awful gods" (*Theogony* 759f.). A mortal trying to enter this region "would not reach the floor until a whole year had reached its end, but cruel blast upon blast would carry him this way or that" (*Theogony* 741), so that he would be *nowhere*, sightless, completely stunned, and without contact with things. And he would be nowhere because he would be within the great *Chasm*, identical with Chaos, which being a void lacks all specification and location. The similarities with Aristotle's scientifically more refined external world, which in memory of the earlier Chaos itself lacks even *void*, cannot go unnoticed.

We do have to take into consideration the fact that dreams are not separate from the rest of the physical world.[68] They are important events in this world. Humans are embedded in the aggregate world, though in a special manner and location. Thus, an objective-seeming report about unusual events could be a report of the kind that is later set apart from the physical world as a dream report or a report about the travels of an independent soul. In Hesiod the relation of the underworld to dreams is certainly very close – we are in the dwellings of the "children of dark night, sleep and death" – that is, the underworld has no place in our waking world or at least in the world of everyday events. Anaximander overcame this state by *driving the nowhere out of the inner core of the cosmos* and banning it to the periphery, which in turn is still surrounded by the *apeiron*. The ban is implicit in his reference to the "other side." This reference sums up a revolutionary change in our world experience in just a brief phrase. From then onward there no longer were any regions, facts, lacunas in the physical world surrounding us humans that would be difficult to reach or subject to special laws. Rather, our *experience* of such places was nothing but a subjective error. This prepared the way for the later general philosophical distinction between real world and deceptive perception.

---

[68] See above, section [16] [*chapter 3.2*].

We might be tempted to argue against Hesiod using *geometrical* considerations – just as Archytas did regarding the world's finitude – based on, for example, the trivial observation that a disk certainly must have a second side. The answer is simple: if the above analysis is correct then Hesiod's world is not accurately represented by geometry.

As far beneath the earth as heaven is above earth; for so far is it from earth to Tartarus. For a brazen anvil falling from heaven nine nights and days would reach the earth upon the tenth: and again, a brazen anvil falling from earth nine nights and days would reach Tartarus upon the tenth.

(Theogony, 720–6)[69]

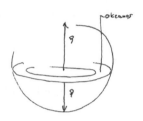

32. Misleading graphic of Hesiod's cosmos

Drawing by Feyerabend.

A scheme such as that in figure 32 (for *Theogony* 720ff. as well as *Iliad* VIII, 15ff.: "[F]ar, far away, where is the deepest gulf beneath the earth, the gates whereof are of iron and the threshold of bronze, as far beneath Hades as heaven is above earth") has as little relation to Hesiod as the common scheme of the Aristotelian world does. Nor should we regard the impossibility of a geometric representation as a valid objection against Hesiod (Hesiod's universe simply is "irrational"). De Sitter's universe cannot be drawn on paper either, and quantum theory gave up long ago on the demand for imageability, that is, amenability to being represented geometrically in three-dimensional space. Paper geometry and its three-dimensional extension, Euclidean geometry, simply *has its limits* both in science and in myths. Attempts to force the world into its categories, which were prepared in Anaximander's work and later became the basis

[69] [*Editors' note: Feyerabend did not actually quote these verses from the Theogony; instead, he merely referred to them. Since, however, the misleading graphic representation of Hesiod's cosmos in figure 32 is hardly intelligible without these lines we have added them here.*]

of astronomy and physics, certainly cause drastic changes. But we do not have proof that all of these changes are advantageous. Let us therefore take a closer look at this situation!

Anaximander's universe is uniform in that even the smallest and most remote elements are subject to the same categories. There are no chasms or lacunas that would require us to switch to another form of representation or to use vague and unspecified descriptions, or where the form of representation that is used has only metaphorical force. Every angle of the model can be enlarged at will and described at any level of detail. Once the perimeter is known it provides a stable framework to accommodate new information. *A spatiotemporal arrangement in a three-dimensional cosmos with complex development trajectories replaces the earlier catalogs.* This gives a new direction to both history and cartography. Just compare Hecataeus' map (as reconstructed according to Herodotus – figures 33 and 34) and its medieval successors (the so-called T-O maps – figure 35) with the Itineraries (figure 38), on the one hand, and the Babylonian world map (figures 36 and 37), on the other.

The Itineraries are just like catalogs in that they list location after location (figure 38), whereas Hecataeus entered the new travel material into Anaximander's scheme and attempted to determine natural boundaries such as rivers, oceans, relative locations of places as well as their positions within the whole – that is, on the circular surface of Anaximander's Earth (figures 33, 34) – by means of geographic directions. Though this led to distortions (just as with regard to the sky, additional symmetrical arrangements were used as well), it did take into account one element that was missing in the catalogs: "A method that had proceeded from the inner core to the periphery instead of the other way around, and that had attempted to establish links between different locations without any constraints imposed by an external framework would have been able to achieve results like the so-called Itineraries for a long time, as the ancient periploi and route descriptions have shown" (Fritz 1967: 64). For Hecataeus even history consists in the attempt to fill in, one by one, those parts of Anaximander's overall cosmic picture that are related to humans (Fritz 1967: 48).

Thus, Anaximander's universe and its later modifications are uniform. *And yet they are not complete.* They do not contain mythological events, dreams or spawns of imagination, since these do not take place at the locations recorded on the map or, even more comprehensively, in the model of the world as a whole. Even categorically *possible* events are rejected as unlikely (Hecataeus' method). And yet

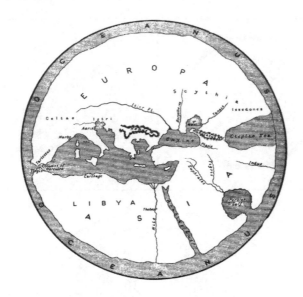

33. Hecataeus' world map

From: Sarton, George (1959): *A History of Science I: Ancient Science Through the Golden Age of Greece*, Cambridge, p. 186.

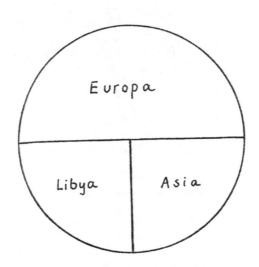

34. Scheme of Hecataeus' world map

From: Krafft, Fritz (1971): *Geschichte der Naturwissenschaft*, Freiburg, p. 171.

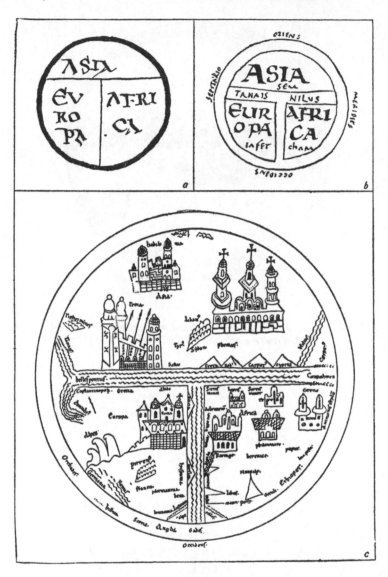

35. Medieval T-O maps

From: Wright, John K. (1925): *The Geographical Lore of the Time of the Crusades: A Study in the History of Medieval Science and Tradition in Western Europe*, New York, p. 67, fig. 1.

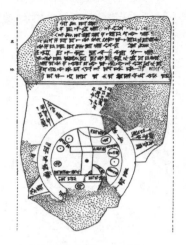

36. Babylonian world map

From: Sarton, George (1959): *A History of Science I: Ancient Science Through the Golden Age of Greece*, Cambridge, p. 84.

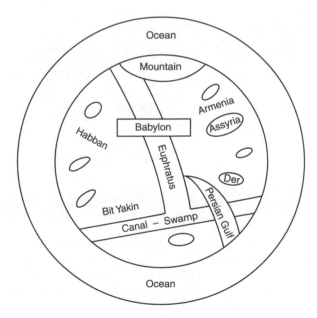

37. Scheme of Babylonian world map

From: Krafft, Fritz (1971): *Geschichte der Naturwissenschaft*, Freiburg, p. 165.

38. Peutinger's itineraries of France and Spain

From: Castorius and Konrad Miller (1887/8): *Die Peutingersche Tafel*. Reprint of
the last revised edition by Konrad Miller including his reconstruction of the lost first
segment with color rendition of the table as well as brief exposition and 18
cardboard sketches of the traditional Roman travel routes for all countries, Stuttgart
1962.

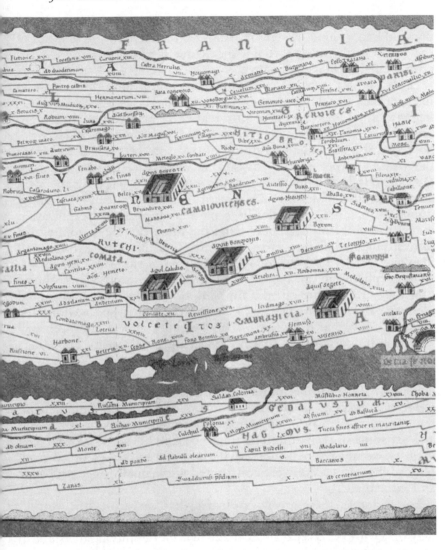

these events have an enormous impact on human life. If we want to take them into account then we seem to be forced to add back some heterogeneous elements into our "uniform" world, after all. This can be done in several different ways.

(1) We integrate the events into the *physical* framework and create a suitable position for them there. This method appears to be at the basis of various travels of the soul to Heaven, which were given a background of objective astronomical facts (Bousset 1960).

131

Plato (*Phaedo* 111e4ff.) clearly proceeded this way: the souls of the damned live within a system of canals that stretches from one end of the world to the other through the center of the Earth and is filled with oscillating masses of water.

(2) We push the events in question out of our physical world and into another world, which is conceived of either as a world consisting entirely of appearance, of non-reality, or as a very real yet non-physical world. We encounter the first alternative in Parmenides, who, however, also relegated everything physical in the standard sense to the world of appearance. The second alternative can be found in Descartes. In both cases we obtain an antiseptic real (physical) world at the expense of insurmountable problems: how can we reunify domains that are so radically separate? Increased specialization makes the problems disappear from the researchers' horizon, which only strengthens said researchers' conviction that their professional ideology is indeed capable of successfully overcoming *all* difficulties. A more philosophical approach reintroduces the problems, shows that they are nothing but the flip side of that revered professional ideology, and thereby gives that ideology the limited space within the realm of ideas that it deserves. A more philosophical approach will also look for cosmologies that no longer introduce such clear breaks into the world.

(3) Cosmologies of this kind do not have to be conjured up; dozens of them can be found in existing ethnological and anthropological research. At all times humans have tried to comprehend the events in this world, and we won't be wrong if we assume that at least some of their solutions might be helpful for *us* as well. (As long as we always remember that the great successes of Western philosophy and science were confronted with equally huge problems and difficulties.)

For example, there are cosmologies that carve up "the world" (that is, the totality of events, as opposed to the materialistic stage play of science) into separate domains without bothering to establish any direct contrast between these domains. There are very abstract, general ideas that can materialize in one form or another depending on the circumstances. The Hopi distinguish between manifest (objective, spatiotemporal, and qualitatively determined) things and things that have yet to be manifested, and the latter are said to derive from the former (Whorf 1956: 57–67). The things that have yet to be manifested are the subjective things. They include the future (which is not a temporal entity but rather a sea of not-yet-realized possibilities), thinking, feeling, wishing, and remote things (which thus do not

132

lie around motionless but become *objectified* only in the course of various activities). The subjective things are not in space though they can turn into spatial things. We may imagine that the indeterminate nature of Hades in Hesiod is due not to a lack of knowledge (of already existing circumstances) but rather to a lack of conditions (needed to realize such circumstances) – it's just that tradition was not very clear about it.

(4) It is clear enough in the early stages of a child's spatial perception. According to Piaget (1954: ch. II) localization at an early age is linked to practical actions, and beyond the realm of practical accessibility the situation fades into indeterminacy (which does not exclude adventurous ideas). Myths differ from the child's situation (or that of an ancient explorer of foreign lands) in that the former conceptually articulate the indeterminate aspects in such a way as to combine them with the determinate ones to form a higher-level unit.

(5) Analogous observations can be made with regard to the idea of complementarity, which was invented explicitly in order to solve some of the problems mentioned in (2) (together with some more specialized problems of microphysics).

(6) None of the aforementioned ideas pays much attention to the events described in *poetry*, which were originally regarded as historical or cosmological *facts*. The structure of the physical universe or of "reality" has pushed poetry further and further into the realm of *entertainment*. A new physics would have to reintroduce poetry as a tool to explore reality.

Now we can also see the actual core of Archytas' "argument." Archytas did not draw our attention to a phenomenon that was possible within the challenged cosmology and had just been overlooked up to that point. Rather, he introduced a worldview *in which things were possible that could not even be thought until then*, and drew conclusions from their possibility. The worldview is not explicitly formulated. It is embedded in an objection that seems to be very simple, challenging opponents to provide an equally simple response. Once opponents acknowledge the objection the game has been won. And if they defend themselves, that shows that they accept the objection as meaningful, thereby conceding again that the game is won, since the objection can be meaningful only if we adopt the new worldview (of unlimited space that is accessible everywhere). Thus, *this* procedure has only the appearance of an argument. Calling it a propaganda maneuver in the form of an argument that presupposes as given what is to be shown, while keeping unsuspecting opponents in the dark about what is going on, would be more accurate. The history of ideas is full of such "arguments" and appears to be "rational" only because philosophers, historians of ideas, and professional rationalists routinely fall for the tricks of their predecessors.[70] Thus, to sum up, we can say that despite numerous points of contact the development from Hesiod to Anaximander was accompanied by a fundamental change in our world experience. And Xenophanes provided the concept of *knowledge* associated with this new world experience.

5.2. Xenophanes: Critic of Religion and Epistemologist[71]

[25] Xenophanes represents the delightful type of thinker who is rarely admired by "serious thinkers." Aristotle called him "somewhat too crude" (agroikoteroi – Metaphysics 986b27), and modern authors regularly place him alongside Parmenides as the latter's "inferior likeness" (Fränkel 1960: 120) or consider him "a thinker of far less sophistication" than even Melissus (Guthrie 1962: 370). He was a much-traveled artist and dabbler, knew his Homer and performed it, and had a keen interest in the new developments in art, poetry (lyric), politics, economy (the invention and impact of money), technology (the Tunnel of Eupalinus, Mandrocles' bridge, Theodorus' Artemision, and revolutionary improvements in naval architecture), cosmology and astronomy (the acquisition and improvement of Babylonian science, Ionian "naturalism"), in the general value system (the gradual replacement of the heroic ethos with the ethos of the city state); he knew about the expeditions and wrote about them in versified aphorisms. These aphorisms were often derisive – indeed, we do not always know whether a slightly eccentric idea presented by him is what he believes in or what he ridicules by exaggerating it – but they also contained sincere admonitions and critical comments on current topics. The admonitions reveal arrogance and anticipate the later institutionalized tyranny of intellectuals. More than ever, human beings were now regarded as the creators of novel things, and they are criticized if they fail to use their intrinsic powers but either remain stuck in old traditions

---

[70] A more recent, though not as intelligent, example of Archytas' method is Popper's "arguments" against the notion of complementarity (Popper 1967). In his objections Popper presupposes classical mechanical materialism without making any effort to consider the arguments that convinced Bohr to give up this philosophy. You'll find a detailed critique of his method in Feyerabend (1968; 1969a: esp. section 8).

[71] [*Editors' note: The larger part of section [25] on Xenophanes (typescript pages 183–98) was missing in Helmut Spinner's version, which otherwise is generally more comprehensive, but it was included in the Constance Archive version.*]

134

("There would be little benefit to the city from this, if someone should compete and be victorious at the banks of Pisa. For this does not fatten the treasury chambers of the city" – DK 21B2) or degenerate into new ones ("While they were still free from the loathsome tyranny, after having learned from the Lydians any kind of useless refinement, they used to come to the place of assembly wearing garments made all of purple [. . .], with boastful mien, delighting in their elaborate coiffure with locks, drenched with ointments of refined fragrance" – DK 21B3). Just like the tragedians, Xenophanes confronted the heroic epic with the new bourgeois morality and condescendingly compared the "ancient fictions" (DK 21B1) with the "useful" (DK 21B1.23) "knowledge" (DK 21B2.12) of the present. As Fränkel (1960: 341) wrote, "A Greek of the sixth century, he dared to dismiss myths as an ancient invention." Among the Ionians he was alone in judging epic ideas so harshly. Humans were no longer embedded in a "meshwork of occurrences" (Fränkel 1960: 14); rather, they *actively* worked to increase their knowledge and change their environment. This was a matter of historical development:

The archaic worldview considered time to be active in the things we encounter: it was the force that introduces everything to us, like a wind blowing toward us the things that happen. For this reason it was always considered "coming," as Pindar liked to put it, that is, as a later time and future. Now, by contrast, in the classic view time is also with us, the ones who experience what happens; it is also in our walk when we approach and walk through the occurrences. Thus, in Aeschylus' tragedy Clytemnestra shows us how she speaks of her bad dreams, "in which I behold more disasters to you than the time of sleep could have compassed" (Aeschylus, *Agamemnon* 894). Thus, time sleeps with the sleeper. Such a shift of time into the person experiencing it is part of the general radical transformation of consciousness that starts with the classical epoch and eventually replaces the ancient frame of mind entirely. The forces active within the events are no longer perceived in the things but drawn into the perceivers themselves. Thus, love is no longer a charm emanating from the loved object in an irresistible manner but a force inside the lover. Actions in general are transformed from mere reactions to occurrences to spontaneous acts of the will. Likewise time is now located within being itself, and no longer just in the occurrences we encounter.

(Fränkel 1960: 13) [72]

---

[72] The older concept of time can be found in Thales' alleged response to the question of

Without doubt this development was substantially promoted by the pessimism of seventh-century Ionian poets. They fully enjoyed the wealth, luxury, and leisure that they received due to constantly expanding trade. Warrior virtues, delicious meals, love, and wine were sufficient for them to make life worth living. Humans were able to obtain these advantages on their own without the help of the gods. Though the gods could have spared the mortals the horrors of aging and death (as Mimnermus put it, "when wearisome age comes upon him, old age which makes a man mean, makes a man wicked as well, evil cares always encircle his heart and his mind and oppress him, nor is he pleased when he looks up at the beams of the sun, since he is hateful to boys and likewise despised by the women"), out of negligence they failed to do so, as they were not interested in our fate. This is the tenor of the poetry of Mimnermus, of the early Callinus, of Semonides, and of the much-traveled Simonides of Ceos. Xenophanes, who like a curious journalist was able to "read the signs of his times" (Edelstein 1967: 14) without simply following them, and turned them instead into new expectations, new descriptions, and a *positive* worldview, put it this way: "From the beginning the gods did not reveal all things to mortals; but, over time, the ones who search will discover what is better" – the ones who search, meaning especially the ones who *think*: for them "the virtue and power of men are manifested in the life of the city, the life of peace, the life of the intellect, of which Homer is neglectful or not yet aware" (Edelstein 1967: 8). What are the ideas that this thinker produced? And what did he have to say about their validity?

In Anaximander and his predecessors epistemological problems arose only in an indirect manner. As we saw, even in Hesiod truth was already very abstract. It was strictly distinguished from deception and lies, and it is the result of a *decision*; that is to say, *the same* abilities that enable us to produce truths can also produce falsehoods. Nevertheless truth was still conceived of as a *sum* of individual reports, a *list*. The lists were never entirely without coherence. Even the old Sumerian catalogs of gods, which contain only words instead of complete sentences, are divided into groups by means of determinatives, and similarly built names were placed in close proximity to one another. The level of inner coherence increased in the task catalogs of Babylonian mathematics, which were held together by principles

who was the wisest person: "Time; for it has found out some things already, it will find out the rest in due time." (Plutarch, *Dinner* 153d).

that admittedly were not explicitly formulated (Soden 1965: 37, 51, 75ff.). In the Western world such principles were included among principles of conservation, which had some significance already in Hesiod and gained more prominence in Anaximander's writings. The fundamental qualities thereby established not only connect later to earlier events, they also combine the chunks of truth contained in traditional representations into a coherent whole, from which the individual is gradually separated due to its inner, more agile subjectivity. This is the core of the later distinction between the one truth and the many opinions held by people, as well as of the corresponding distinction between an observed object and its experience by various observers. Both distinctions, along with their rationales, reveal themselves very clearly in Xenophanes. I shall start with the new notion of *knowledge* that Xenophanes introduced into philosophy and shall then discuss the role of the notion of experience in connection with the earlier notion.

The notion of knowledge that Xenophanes encountered was that of knowledge through one's own intuition (or through the testimony of others). This notion comprises everything that can be said, including mathematics: the word *deiknumi*, which in Euclid has the meaning of "to prove," previously meant roughly "to make something visible," and the earliest "proofs" in mathematics were indeed proofs that would illustrate a complex subject matter on the basis of some simple, perceivable relations (Szabó 1969b: 185–99). Even the gods' messages did not in principle leave the domain of intuitive experience, even though they were about material that was *in practice* barely or not at all accessible to mortals. Xenophanes elaborated on this idea and outlined its limits. Complete knowledge (*saphēs* – "the exact" according to Diels' German translation) combined a correct general description of an object (as in, "This is a table") with a list of all of its details. It is not *given* to us humans with regard either to the gods or to things in nature, nor is it *accessible* to us due to our own nature. Though we can say and think things that express "what really exists," such talk is based neither on direct experience nor on the testimony of others, and so it does not constitute knowledge in the proper sense of the word. *Yet it is valuable nonetheless*, for it can be improved; hence it can be used as if it were already knowledge. This is the content of the much-discussed fragments DK 21B18, 38, 35.

We may be tempted to draw the conclusion that here, as already in Homer, the *huge number* of required experiences would prevent us from ever achieving knowledge in the proper sense of the word: though there are familiar *chunks* of knowledge, any interesting

137

question requires so many experiences to establish an answer that any such answer would lie beyond the domain of genuine knowledge. This is contradicted by B38: "If god had not made green honey, they would have said that figs are much sweeter." Two ideas are combined in this fragment. The first is that we can attribute different qualities to a thing depending on its circumstances. If we combine this idea with another one that Xenophanes appears to have employed freely, namely, that of *reductio ad absurdum*, then this leads us in a very natural manner to a separation of (objective) quality and something *else*: figs cannot be at the same time sweet and not so sweet, hence the change has to be shifted to *another* location. Now if we consider that with the "enlightenment" of the classical era qualities gradually receive a subjective side – "love is no longer a charm emanating from the loved object in an irresistible manner but a force inside the lover"; sweetness is no longer only an external quality but also an inner sensation – then it becomes natural to separate this "other" location as "sensation" from the real object itself. DK 21B38 would then mean that the "evidence" available to us changes in different circumstances, while the "real process" remains the same and therefore requires interpretation. Such interpretation has to take *all* circumstances into account; that is, it requires knowledge that is generally not accessible to humans. With that, not only does a lengthy discourse reveal itself to be unreliable, *but intuition itself* does so, and "belief is fashioned over all things"* according to Diels B34's arguably correct translation. The bottom line is that the discourse associated with empirical observation *consistently gives only the appearance of truth*. (We shall look at this result from another angle later on.)

The discourse associated with empirical observation consistently gives only the appearance of truth. *But that doesn't matter*. Just as humans can build and improve bridges without divine assistance, so can they build and improve true-appearing discourses for human needs. Though this is not "knowledge" it is an important part of civilization and furthermore for the first time clearly distinguished from the objects it is about: not only does Xenophanes' theory give human knowledge the place it deserves within culture as a whole, but it also *distinguishes* this knowledge more clearly from nature and from the rest of culture than had been done before, which is why this is the first actual *theory of knowledge*.

---

* [*Translator's note: this is the standard English translation of the phrase following Richard D. McKirahan, Philosophy Before Socrates, London, 1994.*]

Hermann Fränkel believed that the seemingly skeptical ending of DK 21B34 ("belief is fashioned over all things" or "belief is fashioned in the case of all persons") could not be reconciled with Xenophanes' robust realism, his practical attitudes, or his life (Fränkel 1960: 342ff.), and he unequivocally rejected Sextus Empiricus' version: "For even if a person should in fact say what is absolutely the case, nevertheless he himself does not know" (*Logicians* VII, 51f. = DK 21B34). Reinhardt also experienced difficulties in his attempt to combine Xenophanes' "strong preference for reality in every respect: experience, appearance, detail, rationality, and usefulness" (Reinhardt 1959: 144) with a "tender skepticism" (Reinhardt 1959: 151). The two honest gentlemen[73] appear to have overlooked that Xenophanes may have spoken in a somewhat ironic manner here, for example like this:

> What you call *knowledge*, parrot as an old wisdom, and pass on from one generation to the next, is not accessible to humans. If they have it then they don't have it at the same time; for they do not know if they have it. And yet your ordinary discourse, which is certainly not knowledge, can be improved just as your bridges and ships can be. So be content with it and try to improve *it*, in the same way that you are content with bridges and ships that you have built on your own without the help of the gods.

Fränkel's own translation, roughly "he does not have knowledge from his own experience," and his explanation – "Xenophanes regards only that kind of knowledge as certain and exhaustive (*saphēs*) that is empirical. Thus, he considered only *opsis* and *historie*, to use two of Herodotus' terms, as reliable" (Fränkel 1960: 348) – do not coordinate well with DK 21B34, and also contradict the confidence with which Xenophanes presented his ideas about the notion of gods. If we trace this confidence back to Parmenides then the explanation dissolves into nothingness.

There is yet another location where sarcasm works as at least *one* possible explanation for statements that are not easy to comprehend. The infinite depth of Earth either is a sarcastic reference to Anaximander ("If you can circumvent the question of genesis by declaring the *apeiron* to be infinitely old then I can circumvent the question of the Earth's fundament by declaring it to be of infinite depth") or should be interpreted as a reflection of the nature of our knowledge: we know the surface of the Earth well, but upward and downward from there our knowledge becomes indeterminate. We have to be at a certain location to be able to say anything specific

---

[73] Fränkel wrote about Reinhardt that the latter "had grasped Parmenides' teachings in all their depth and understood them" (Fränkel 1960: 157 n. 1).

about it. Some authors have noticed the novel, sarcastic nature of Xenophanes' writings. Gigon writes: "The inconstancy of the standpoint characteristic of elegy since Archilochus must have bestowed a special nature on Xenophanes' philosophy as well" (Gigon 1968: 157), and "this assumption [of several suns] is so grossly empiricist that the idea suggests itself of Xenophanes being less concerned here about an objective meteorological theory than about polemically exaggerating rejection of the Milesian construction. Not a few of his interpretations must have had an Archilochean coloring" (Gigon 1968: 169). And even Reinhardt assessed DK 21B7 as follows: "The howling of dogs as the familiar voice of a friend, the deep-rooted compassion, and the entire sparsely and yet so fittingly caricatured sublimity of the great wonder man [Pythagoras] are unsurpassable" (Reinhardt 1959: 141). Would not the same mockery also have left traces here and there in Xenophanes' didactic poetry?

The separation of knowledge from what is known was by no means a matter of course during the archaic period. The poet of Homer's shield description (*Iliad* XVIII, 478–607) views the things themselves through the picture as if through glass – he does not appear to be capable of grasping the picture *as* a picture. Concepts that express knowledge directly describe the relation between humans and their environment without distinguishing the latter from special processes within the human observer (Guthrie 1965: 17ff.), and even in Parmenides' fragments "a separation between thinking and being [. . .] simply cannot be performed" (Reinhardt 1959: 30). But in Xenophanes the distinction is clearly made.

If the cultural element of true-seeming speech was to fully enter human possession then everything had to be removed that presupposed divine ideas with regard to either content or origin. This "knowledge" was to be built out of *human material alone*, of course with the exception of those ideas that occur in "ancient fictions." This is what determines the structure of Xenophanes' "cosmos." This world does not have a heaven, since the notion of heaven is too closely connected with theological notions.[74] The world extends upward into an indeterminate vastness and downward in the same manner; the upper part is the air's dominion and the lower part the earth's. Humans are at the intersection of air and earth; they are born of earth and water, which is the source of rivers, rainfalls, and clouds, and which periodically floods the Earth. Xenophanes

---

[74] This material is from Guthrie (1951: ch. VIII).

appeared to have found evidence for floods in fossilized fish and marine plants in Syracuse, Malta, and Paros, as well as in the water dripping down from cave walls. Much like clouds, a new sun is formed each day; it runs in a straight line above the surface of the Earth and disappears in the west. It does not "rise" or "go down," since these notions, which have little or no support from experience, are remnants from the ancient mythical past. Just as every country has its own clouds, so does every country have its own sun, and so forth.

Ideas such as these earned Xenophanes the reputation of a plain and unimaginative empiricist. "This theory seems oddly primitive and violent even compared to the times of its inventor. It is poorly thought out and of inferior quality," wrote Hermann Fränkel (1960: 340). "Overall the strange man comes across as rather unphilosophical" (Fränkel 1960: 339). And Karl Reinhardt, as always in search of genuine depth, evaluates Xenophanes as follows: "With respect to an explanation of the world Xenophanes appears as a philosophizing dabbler who always goes with the next-best, crudest solutions, incapable of going into depth regarding any problem" (Reinhardt 1959: 145). The same author wrote about Xenophanes' denial of a heaven: "Here Xenophanes dared to do what no other Greek philosopher since Thales had dared: he managed to flatly deny the spherical shape of the sky and the world and thus the connections between the stars' motions" (Reinhardt 1959: 146). Now, first, it is not certain that the notion of a spherically shaped heaven dates back to Thales. On the contrary, there is no evidence for this. The notion is missing in Homer, apart from a few isolated passages that are not very clear; and we find it in Hesiod due to the latter's mythical approach (separation of Heaven and Earth), though there it is not directly linked to the locations of the stars. Anaximander was the first to fix the stars temporarily in their rings but was then forced to set certain of these rings in motion in a precisely determinable manner. He also combined traditional ideas, principles of symmetry of which he was only semiconscious, original insights, and occasional observations into a whole in which all of these components are in seamless transition one to another. Was it "all right" to dissolve this whole and, in particular, was it "all right" to remove from it the notion of heavenly rings and thereby the notion of a well-built Heaven?

From our contemporary point of view the answer is clear. *Methodology* teaches us that the most thorough improvement of a theory requires the use of alternatives negating essential components within the theory. *Astronomy* teaches us that there is no "Heaven,"

141

that the fixed stars "move about free and unbound" (Galileo), and that the planets are held in their complicated orbits not by spheres but by forces. Xenophanes' cosmology is consistent with both of these notions, and in this respect it is entirely "modern." But even in its own historical environment its underlying radical empiricism appears "naïve" and "unphilosophical" only if we assume that Xenophanes used *already pre-existing* empirical ideas and that he was content with "naïve and unphilosophical" versions of them. This assumption is at the root of all of the condescending assessments of Xenophanes' accomplishments, and it is false. It is correct that even long before Xenophanes appearance played an important part in archaic thought. *However, appearance and experience in the empiricist sense were two different kettles of fish.* Experience stood in contrast to untestable and erroneous *speculation,* which is lacking an empirical basis and incapable of having one, and to the "ancient fictions," as Xenophanes condescendingly calls the myths (DK 21B1.22). Furthermore, experience stood in contrast to *sensory appearance,* from which no conclusions about the environment can be drawn. Appearance, however, also stood in contrast to rumors, to witness testimony, and to messages from gods, all of which can be traced back to it in principle, *and it also confirms the "ancient fictions," which experience denies.*[75] Yes, it is hard to imagine anything that would not be accessible to appearance, though at times it may be only in special and hard-to-realize circumstances. The step from appearance to experience limited this resource and conveyed the revolutionary admission that something we see clearly does not, after all, have to be the way we see it (only a little later there would be a complete denial that we have seen anything at all – this is the basic assumption of the naïve theory of nature myths). Furthermore, it now became necessary to find a new criterion of existence that would go alongside appearance *and occasionally contradict it.* It is likely that Xenophanes made this step, though not entirely without outside help. If this conjecture is correct then he was the inventor not only of epistemology but also of empiricism, as well as the first thinker who knew how to use dialectic thought procedures elegantly and with a light hand.[76] So let us see how this conjecture can be supported!

We know that Xenophanes recognized traditional notions of gods and the customs associated with them as relative and that for this

[75] On this topic see above, section [16] [*chapter 3.2*], including the comments. There I speak of "experience" rather than of appearance in order to suggest that modern empiricism has blurred the distinction again.

[76] On the latter topic see Reinhardt (1959: 104).

reason he classified them as mere human fictions (DK 21B15, 16). Such criticism was not new; it was widespread in the enlightened circles of his time. Herodotus, for example, told the following anecdote about Darius (*Histories* III, 38). Darius asked the Greeks at his court whether they were willing to dine on the corpses of their parents. Their answer was "not for all the money in the world"! He then asked "Indians belonging to the tribe of the Callatians," whose custom was to eat the corpses of their parents, for how much money they would burn the corpses. They let out a cry of dismay and begged him never again to mention something so callous. Herodotus wraps up his anecdote with "thus we see the power of customs; and Pindar, I believe, was right in calling them the 'king of all.'" This "king of all" did not just accommodate reports about gods and expectations directed to humans. Closely associated with both were cosmological ideas – such as Earth's being completely covered by Heaven – that were repeated in only slightly altered clothing in contemporary philosophical speculations on nature. To conduct a reform of traditional notions of divinity and replace them with "a superior and purified conception of the divine" (Fritz 1971: 36; see also Reinhardt 1959: 99) – and everyone welcomed this reformation by Xenophanes! – we cannot avoid a reformation of cosmology as well.

Such a reformation has to be first and foremost a reformation of *concepts*. For archaic thought does not sharply distinguish between cosmological concepts and theological ones. We saw that Uranus and Gaia are hermaphrodites displaying at times a personal aspect and at others a purely material one without ever entirely losing either.[77] The unit comprising both aspects is not easy to understand from today's perspective. We are tempted to think that such a unit could not be anything more than a highly artificial and easily dissolvable conglomerate. This is correct if we apply modern concepts of divinity and matter (though there are exceptions even here). Archaic times were not yet familiar with such antiseptic entities, and the diversely enigmatic processes in the world were fittingly described by means of equally enigmatic speech. After all, the two aspects of Uranus and Gaia were not just conceptual constructs; rather, they were components of contemporary experience. Both for Thales and for the farmers of Boeotia all things were in motion and filled with gods. If one denied the gods then one thereby also denied the conclusiveness of evidence for the material aspects of the world, unless one had

---

[77] For similar concepts in Near Eastern thought see the comments on "You" and "It" in section [10] [*chapter 2.1*].

dissolved the close connection between the aspects and created new concepts of *divinity* and of *humanity/materiality/reality*.

And this is what Xenophanes actually did, according to all relevant authors: "He deprives the visible heavenly phenomena of all divinity and takes their very dignity and duration away, just as on the other hand he denies his god all comparability to mortal bodies" (Fränkel 1960: 348). "For Xenophanes, due to the transcendence of his god the individual human being, the polis, the nation, and even the natural world are pretty much forsaken by everything supernatural" (Schachermeyr 1966: 45). The tendency here was exactly the same as in the case of technology and knowledge. Not only were human *activities* – such as the activity of multiplying one's knowledge, or that of the improvement of ships, or that of legislature – purified of all divine elements and placed by themselves, but even the *environment* in which these activities take place lost its divine aspect. The divine was no longer to be found in nature, and even dreams – which once were regarded as direct empirical proof of divine intervention – were now explained in purely physical terms (DK 21A51, 52). We may conjecture that Xenophanes' "dilettantish" cosmology and "unphilosophical" empiricism are simply the flip side of his "superior and purified concept of divinity" and hence not quite as "naïve" as they appeared to some modern readers. Let us therefore first examine the reasons for Xenophanes' new god!

Looked at in an unbiased way this god is an unfriendly monster, behind which we will notice with dismay the self-image and increasing lack of tolerance of the newly developed intelligentsia. This god is "not like mortals either in bodily form or in thought," "It is as a whole that he sees, as a whole that he thinks, as a whole that he hears," he "without effort sways all things by the thought of his mind," "always he stays still in the same place, not stirring at all; nor is it fitting for him to move about hither and thither" (DK 21B23, 24, 25, 26). There was no precedent for such a god within the *tradition* (for how would an intelligent and courageous god get the idea to not "move about hither and thither"? And what is the spherical shape that is also attributed to that god about – DK 21A31?). Thus, whence this odd shape? Consider the following piece of historical fiction, which *may* contain a large amount of truth.[78]

The god was introduced because Xenophanes was able to *prove* his most important qualities. These qualities were not determined by

---

[78] [*Editors' note: This sentence was a handwritten later addition to the text; unlike the other parts of the text it was written in English rather than in German.*]

sensible intuition, or by tradition, or by the demands of piety, but by certain intellectual rules of the game that had just been discovered yet already figured as conditions of existence and non-existence. In brief summary, the proofs look like this. God has to be *one*. For if he were many, then the many would have to be either identical or not identical. If identical then he would be, again, one; if not identical then he would not be existent in more than one place, hence he would not be many, after all. He must *not* have been *generated*. For if he had a genesis then he would have had to be generated either from something identical or from something non-identical. If he is generated from something identical, then he is not actually generated; but he cannot be generated from something that is not identical, either, because something that is cannot come from something that is not. A similar proof can be used to establish God's *omnipotence*. An omnipotent god with a genesis derives either from something identical – that is, from something equally omnipotent – or from something non-identical. The first case is not a genesis at all, because he remains the same throughout. In the second case he is generated either from something more powerful or from something less powerful. He cannot have been generated from something more powerful, for if he had been then the more powerful entity would necessarily still exist. Nor could he have been generated from something less powerful – for whence would the less powerful entity obtain the strength to generate something more powerful? In all of these cases we have an indirect proof applied to the dichotomy identical/non-identical, and the argument is of the form: if A then either identical or non-identical. If identical then A cannot be true. But non-identical cannot be true, either; hence A cannot be true. Xenophanes' concept of god was perfectly suited for these kinds of proofs, *which is the reason why Xenophanes selected it*. That this concept was not backed up by popular belief "did not concern him, for only the unit concept was dialectically conceivable and provable for him, and dialectics was what mattered most to him" (Reinhardt 1959: 96).

Here we have for the first time a most interesting phenomenon that today belongs to the basic principles of scientific concept formation. An object is introduced because certain rules of the game can be applied to it and because the application yields exciting new results, and not because we encounter it in sense perception or tradition. In Xenophanes the object thus artificially constructed also superseded the empirically observable gods, thereby declaring essential parts of sense perception to be unreliable. *The playful demands of the intellect were pushed to the forefront* (the playful demands of the intellectual class became more important than our connection with scientific practice or the thoughts of our fellow citizens) *and determined which objects from then on were to be regarded as existing in this world.* The gods were ousted by an arbitrary supposition rather than by proof of their irrationality or ineffectiveness, and replaced by a motionless, emotionless, spherical, and inhuman super-boss. (How could anyone ever get the idea of calling this a "superior and purified concept of God"?) Of course, we need to ask about the origins of the method of proof on which this supposition is based. I shall turn to this topic in the next section ([26] – [*chapter 5.3*]). In the present section I just explore the *consequences* of the procedure.

Thus, a new, abstract concept of God that can be grasped only in thought replaced the traditional concept of gods. This was not a simple process. Gods tend to appear to us in dreams, they are seen in broad daylight, and we can feel their force in our own bodies and thoughts.[79] Divine forces are active in nature, and gods gave us humans the arts and sciences. Removing them means having to devise a new concept of knowledge, a new approach to politics and technology, and a new description of nature, or in other words: we need new concepts still unspoiled by the "ancient fictions," a cosmology that solely relies on what is accessible to humans, and in particular, we have to revise our *sensible intuition*, which had until now connected humans with the god-animated nature surrounding us. Thus, we essentially also need new humans who either no longer notice gods, living phenomenally in a godless kind of nature, or who no longer pay much attention to those intuited appearances of gods that remain. These are the tasks that Xenophanes faced, and he completed them all. I have already described the new, pragmatic approach to politics and technology that emerges in his work, as well as the new concept of human knowledge that corresponds to this pragmatism. This change in concepts belonging to philosophy of nature resulted in the first *materialistic* theory of the world, and the change (restriction and/or reinterpretation) of intuitions makes him the inventor of *empiricism*. The concrete implementation of his empiricism (the sun as a cloud) is barely less rational than the modern notion of an object as a bundle of sense data; on the contrary, it has a much stronger theoretical foundation. Thus, Xenophanes' "dilettantish" lore turns out to be a natural, though not at all dilettantish, consequence of his

---

[79] See again the exposition in section [16] [*chapter 3.2*].

abstract concept of God.[80] And the latter concept itself in turn paved the way for *Parmenides'* philosophy.

The development of empiricism reconstructed here shows that arguments had no part in its birth. Instead, certain rules of the game were introduced to handle concepts, a concept was selected that would lead to interesting results if those rules were applied to it, and that concept was used to replace the familiar concepts of gods, which had strong support both in tradition and in sensible intuition (experience in the broader sense); then religion, cosmology, and intuition were emasculated, which was eventually presented as the invention of a new method. The trick would have been acceptable and even commendable if it had resulted in an improvement of the situation. But why should we regard Xenophanes' monster as "purer and more refined" than, for example, the Thracians' blue-eyed, red-haired gods (DK 21B16)? Does a being become more refined by losing human traits and acquiring inhuman ones? And in what way is the experience of a junk room – for this is, after all, what Xenophanes' world resembles – superior to that of a world animated with gods? It seems that viewing Xenophanes' admittedly quite quirky and intelligent step as *progress* would require the one-sided and very narrow approach of today's science nuts.

5.3. Parmenides: The Origins of Western Philosophy of Nature

[26] A look at *Parmenides'* philosophy shows that Xenophanes' concept of God is a *composite* of two distinct yet converging notions. First, there is the gradually growing conviction that a single god is actually behind the diverse god figures of the epics: "Often my mother Themis, or Earth (though one form, she had many names)," wrote

---

[80] This interpretation of Xenophanes is based on a piece of writing called *On Melissus, Xenophanes, and Gorgias*, which has been included in Aristotle's work but which can hardly have been composed prior to the first pre-Christian century. Today some scholars challenge the reliability of the document, but their arguments are not very convincing. For example, Reinhardt's remark that the document is essentially accurate about Melissus and that it therefore should be trusted also with regard to Xenophanes (Reinhardt 1959: 90ff.) is met by Guthrie (1962: 370) with the response that the author as a student of the Eleatic tradition would naturally do justice to its representatives, which does not necessarily mean that the author's report on Xenophanes is equally trustworthy. First, this reply overlooks the fact that Reinhardt also checked the Gorgias part of the document (with the result that the anonymous author is more reliable than Sextus Empiricus), and second, it also overlooks that the document presents Xenophanes as an Eleatic himself, so that Guthrie's own argument would actually compel us to trust its report on Xenophanes. See also Gigon's assessment (1968: 192).

147

Aeschylus (*Prometheus* 210). This conviction is linked to the very abstract and completely impersonal proofs of the unity, simplicity, indivisibility, and invariability of being that we find in Parmenides.

My account of Parmenides will have to be even sketchier than that of his predecessors. Everything about him is in question: the order of his fragments, the details of his writing style, and his translation, interpretation, and historical classification. The best we can do in such circumstances is to tell a *fairytale* only loosely linked to the accepted facts (for a *firmer* connection would attribute more power to the facts than they actually possess), one that connects events in an interesting and stimulating manner. The objection that fairytales have no place in history or philosophy can be met with a simple response: *any* historical account that does not restrict itself to mere catalogs, as Sumerian and Babylonian history do, transforms the material at will while rarely being aware of such transformative activity.[81] Thus, let us explore what stimulating possibilities we can find in the Parmenides material!

Parmenides' aforementioned "proof," which completely transformed philosophy of nature, is the first explicit proof of a principle of conservation in the history of Western thought. This is not as anachronistic as it sounds. "Principles of conservation" had been prepared in cosmology since Hesiod, and we have reason to conjecture that they played an important part even in "primitive" thought. Since Parmenides did not have a clear distinction between concept and object, he was compelled to demonstrate directly in the *things* themselves what we today can find out by analyzing *sentences* and their relations to one another. (Since every semantic analysis presupposes that the analyzed entities reflect the world more or less adequately, this "clinging to the things themselves" is not quite as naïve as one might think. It is certainly not any more naïve than the purely formal approaches of modern logic, which have a lot in common with linguistic hocus-pocus.) What for us today is divided into cosmology (physics), epistemology, logic, semantics, and so forth was for him still unified as general physics.

Knowledge, for example, is for Parmenides a physical process, and the statement "Like is known by like" suggests not a correspondence of something in nature to an entirely different, subjectively guided thinking, but rather one of substances in nature to *the same* substances in human beings. Here the old wisdom works:

---

[81] See the comments on section [16] [*chapter 3.2*].

148

that knowledge consists in proper adaptation to one's environment (section [16] – [*chapter 3.2*]). Humans can gain knowledge about light, life, heat, and sound to the extent to which light, which is conceived of as a physical substance, is primarily present in them. If it temporarily replaces the dominion of darkness, the second world-substance according to Parmenides, then we have a "condition of total enlightenment that does not leave any room for any earthly remainders within the enlightened person and completely fills them with genuine being" (Fränkel 1960: 179). But if all light is replaced by the dark matter, human beings turn into corpses and as such cannot have "knowledge" of either light or sound, or heat, though they are capable of gaining "knowledge" of that which is like the dark matter, namely darkness, coldness; they also hear the inaudible and feel what is dead (DK 28A46.13ff.). This is the reversal of Odysseus' doctrine: "[F]or the spirit of men upon the earth is even such as the day which the father of gods and men brings upon them" (*Odyssey*, XVIII 136f.), which appears to have found poetic expression as well (Archilochus, *Fragments* 68; Heraclitus, DK 22B17). Circumstances do not shape our thought; rather, our thought – that is, the balance of primordial substances present in us humans – selects our circumstances. The doctrine seems primitive compared to the glacier-covered conceptual labyrinth that we encounter in the *Critique of Pure Reason* or in its modern offshoot, the *Logic of Scientific Discovery*. But let us remember that genetics, biochemistry, neurophysiology, and physical optics are collectively returning to that doctrine, and that there are even purely psychological aspirations to remove humans' alienation from their environment and replace it with a more intimate connection, a new natural integration into the world. In the first case the attempt is to understand the process of collecting and interpreting information on the basis of the same "substances" and laws that we encounter in the external world of matter. Parmenides was the first to formulate the research program at the basis of this endeavor in a simple manner and at the same time take the necessary steps to develop it. Like all other thinkers of the time, in doing so he relied heavily on the ideology of the epics.

This brings us to the third element in Parmenides' investigations. He selected the *most general* quality, namely that which exists, or *Being*, as the basis for his doctrines. With respect to this quality he proves his general *principle of conservation*, "Being is," as well as its *corollaries*, "Being is continuous, homogeneous everywhere, uncreated, imperishable, motionless, one, and spherical." The key claim in the preserved fragments does not have the form *to ōn estin*. We find "*esti gar einai,*

149

mēden d' ouk estin" (DK 28B6.1f.), and we are asked to decide *"estin ē ouk estin"* (DK 28B8.16).[82] Then again we find simply *"estin"* (DK 28B8.2) – "it is" – for the proposed general principle of conservation. The additions are natural in light of the corollaries, all of which concern the subject *estin* of DK 28B8.2. The objection that this turns the basic claim into a tautology is neither compelling – *to ōn estin* does not have the form of a tautology – nor relevant: after all, even *after* Parmenides thinkers still had to respond to Zeno's argument that "half a given time is equal to double that time" (Aristotle, *Physics* 239b33) and to rule out this possibility by means of axioms of equality that strike us today as trivial (Szabó 1969b: 291ff.).

Physical and logical elements are inseparably linked in the arguments for the principle of conservation and its corollaries. Whatever exists possesses, in addition to the special properties that human opinion ascribes to it, the property of Being, which is hard to recognize due to its continuous presence everywhere, and its characteristics are determined specifically by *this* property. Whatever has this basic property is determined by a complete listing of the most general possibilities as well as by an elimination of the impossibilities. It is not direct *intuition* that decides about the basic nature of the world, namely that it is everywhere in the same manner, but rather a consideration that secures this nature via the elimination of entirely non-intuitional elements.

The possibilities (or "ways") under consideration are: "It is," "It is not," and "It is and is not." "Only one remains" (28B8.1f.), namely *estin*, while the two other alternatives are eliminated by means of a simple argument. Since nothing can be known about nothingness, nothing can be said about it in principle (28B2.5ff.), which eliminates the way of nothingness from our consideration. The result is a positive statement even about the world of intuition, so that this world becomes subject not only to the judgment of intuition but also to an additional type of judgment, namely that of thought (28B7.5). It was to be expected that intuition would sooner or later come into conflict with this new type of judgment, thereby losing its absolute status as a source of reliable knowledge. Normal epistemic behavior, which consists in the natural integration of human beings into their environment (section [16] – [*chapter 3.2*]), became subject to a new authority,

---

[82] [*Editors' note: Feyerabend did not translate the Greek text. "to ōn estin" means "Being is"; "esti gar einai, mēden d' ouk estin" means "for Being is, though Nothing is not," and "estin ē ouk estin" means "is or is not."*]

which started with humans and gradually released them from their environment.

For a time the new authority was introduced as a quality of the world bestowed by the goddess, a quality that we can *find* within the world, rather than one that was *imposed* on it. That changed the function of the myth surrounding Parmenides' arguments. Rather than repeating the basic categories of *tradition* in the shape of concrete events it now became the provisional outline of a *new* ideology, *new* organizing principles, and a *new* form of world-intuition. We could also say that in Parmenides myth introduced fragments of an unusual, not yet fully articulated language game, thereby guiding human thought in a different direction. Indeed, the three possibilities are not, after all, elements of an already existing stock of knowledge but rather building blocks for a new one.[83] The contradiction between Parmenides' arguments and traditional intuition becomes clear in the corollaries: Being cannot grow out of Being, since that would not constitute growth in the first place; neither can it grow out of nothing, since the latter does not exist. Similar considerations apply to the notion of perishing. Thus it does not undergo changes, development or degeneration. It is one indivisible unity, "for there is not here and there a stronger Being that could obstruct its unity, or a weaker one" (28B8.23f.). Despite all the intuitive diversity and change the world is, after all, an indivisible and unchangeable whole.

Upon completion of his proofs Parmenides presented his cosmology, which reintroduced with some characteristic alterations the tradition of changing and developing things that had been tossed out in the proofs themselves. *Night* and *light* were the basic opposites that pervade everything. They did not merge like the elements found in popular thought and early thinkers, but were strictly separate from one another on the basis of the refutation of the "third way" (28B6.8f.). Coming into being and perishing were no longer qualitative changes – this had been ruled by the principle of conservation of Being – but rather compositions and decompositions of the basic substances. Even humans and human knowledge were explained purely physically in terms of the composition and composition ratio of basic substances, and thought itself was nothing but a more or less in the composition ratio (28B16.4). Special procedures were required to determine the continuous unity and permanence of Being despite

---

[83] This anticipating function of myth is clearly pronounced in Plato: whenever he reaches the limits of conceptual considerations Plato introduces myths to indicate at least the core of further lines of development.

this dependence of knowledge on change. This would be one interpretation of the cosmogony, about whose intent there has been much
disagreement (Schwabl 1953).

[27] We can properly evaluate the ideas and methods that a thinker
introduces only after observing how they develop over time, and
only by examining the discoveries that result from them. Looked at
in isolation ideas often seem enigmatic, primitive, or naïve. The contemporaries who take them up and develop them further will explain
their meaning within their historical context and make their potential
clear to us. Whenever these contemporaries invent something that is
now familiar to and treasured by us, we are far more willing to give
such an undertaking our blessing.[84] Thus, let us see how Parmenides
affected his contemporaries' and his successors' thinking and what we
can learn from him today!

Right away Parmenides' immediate successors, such as Xenophanes,
Melissus, and Zeno, adopted the new thinking toy and used it in a
much less forced manner. They extracted two elements from the toy,
dichotomy and indirect proof, and applied both in dialectic fashion
to all the qualities of Being mentioned by Parmenides in 28B8. They
developed a love of paradoxes (see Gorgias with his work on Non-
Being) as well as of the increasing restriction of intuition as a source
of knowledge. Instead of letting the environment "guide" them,
they were able to put the results of such "guidance" *as well as the
entire tradition* in their place using their own fascinating activities.
The "seeing together" of the intuitive elements, which the earlier
aggregate approach did not practice but which was now suggested
by the theorem of Being and Non-Being, together with the use of
one approach to the results of diverse integrative processes into the
environment, led to additional contradictions, to that "chaos of
impressions" that was so often mentioned at that time. What contains
this chaos is not intuition itself, but rather the monistic unification of
elements that had previously lived together peacefully side by side but
that came into conflict with one another once they were evaluated
for universal validity. It was only then that the bent oar in the water
became a crucial argument against the usefulness of perception. It

[84] I already noted that Parmenides' theory of humanity anticipated contemporary research
of our time that aims at explaining even human knowledge naturalistically. While the
approach is the same, the details are different. The method of introducing new ideas that
Plato perfected was actually ahead of our contemporary thought. For today we too often
restrict ourselves to a precise and clear presentation of knowledge that has been *already
obtained*, and no one cares about how *fragments* of knowledge and forms of argument
should be handled within a debate. With some exceptions, such as Pólya and Lakatos,
the form of informal thinking is still largely unexplored.

was once again the consequences of a new approach – including a new way of experiencing the world – that led to the separation of object and perception, intuition and thought, aspect and reality, and to all those problems that resulted from the separation.[85] There was no *theoretical* necessity to leave Homer's aggregate universe, and *experience* certainly did not support such a change.

The idea of the unity and invariability of Being also gave rise to a new concept of *elements*. Elements were no longer able to merge, nor could they consume one another. The only change that could take place was the composition and decomposition of unvarying and strictly distinct basic substances. For the atomists this theory was very far away from intuition. Two principles were used: the primordial building blocks, each of which by itself satisfied Parmenides' criteria of unity, and the medium in which the composition and decomposition took place, the Being Non-Being or space. Note the tendency here that is commonplace in contemporary science: the thinker, that is, in antiquity the philosopher, deals with things that do not occur in intuition and that have paradoxical qualities. *These* are the things that are "real" for the thinker. *They* are what he uses to build complex objects whose grasp by humans yields something similar to intuition. The rift was never entirely overcome; rather, the "real world" became ever more remote from the world in which we live and feel. It becomes institutionalized until eventually the power of the growing institutions of science and an education regulated by them closes the rift from the other end by means of a kind of training that keeps transforming intuition, the behavior regulated by it, and thereby us humans as well until we obey the scientific forms of thought and see the world through them, as a junk room deserted by gods yet well organized.

The new concept of element was accompanied by a new *representation* of cosmological truths. In myths and in the work of Ionian philosophers of nature the structure of the world is represented via a narrative relating how it developed. Thus we have *cosmogonies*. But now, since change had turned out to be a fiction, the nature of being is described by using *axioms*. Wherever "axioms turn up in Pre-Socratic philosophy we need to turn our attention first of all to the Eleatics" (Reinhardt 1959: 105). Parmenides initiated this procedure with his axiom *estin*. Myths continue to be used, but only as a preparation, as a makeshift procedure: wherever new ideas are not yet ready for

---

[85] In painting some of these problems were solved by means of perspective, but they were still far from being understood.

axiomatization, myths *teach* us *roughly* the use of basic concepts; furthermore, they also *impress* them on our memory, so that it will be possible to follow up this first step with a second step after some period of time. In this way myths became a means of logical preparation and an instrument for re-education at the same time. The fact that they were also palatable smoothed down the severity of the conceptual procedure a little.

In *mathematics*, too, the use of axioms can be traced back via some detours to Parmenides. This has been shown recently by Árpád Szabó in a number of groundbreaking and highly interesting studies. The oldest system of Greek arithmetic essentially "is nothing more than an extension of the teaching of the Eleatics" (Szabó 1969b: 261). This can be seen in Euclid's definition of a unit (book VII, first definition: "Unit is that by virtue of which each of the things that exist is called one"), which seems irrelevant and vain until we recall that prior to Archimedes Greek arithmetic did not allow for any fractions. The definition is the theoretical justification of this procedure: "It is definitely not the vacuous descriptive statement which it appears to be at first sight. . . . It represents the conclusion of a carefully considered argument and implicitly determines *whether or not the one is divisible*" (Szabó 1969b: 260). As Szabó has compellingly shown, the reason for this determination lies in the Eleatic arguments for the indivisibility of One. "It is fair to say, therefore, that the Euclidean definition of 'unit' is nothing but a concise summary of the Eleatic doctrine of 'Being'" (Szabó 1969b: 261). The special status of the number One in Eleatic-influenced arithmetic would remain prominent for a long time in that arguments would take the form "if *a* is a number, *or the number One* . . . ." Euclid's second definition "A number is a multitude composed of units," however, aims at making arithmetic possible despite the Eleatics' denial of multitudes. Its role is similar to that of *aitemata* (requirements, postulates) in dialectical discussions asking dialogue partners to *concede* a certain assumption for the sake of the argument, even if they are inclined to *reject* it for good reasons. This multiplying of "the one" (Plato, *Republic* 525e) is the arithmetical counterpart to the multiplication of the unit in atomic theory.

In geometry, however, history of mathematics has to solve the following problem. How is it possible that a discipline so closely linked to intuition as geometry was, and which was proud of this connection and its practical achievements (land surveying), holding it in high appreciation and using intuition in crucial locations within a proof (see the discussion in *Meno*) – how is it that such a discipline suddenly turns away from intuition and attempts to proceed in purely abstract terms? Szabó (1969b: 98ff.) traced this turnaround back to the discovery of linear incommensurability. That discovery demonstrated the unreliability of intuition, resulting in the adoption of the method of indirect proof (which was developed by the Eleatics), as well as in the more general question of whether the Eleatic way of thinking could be applied to geometry as a whole. The postulates of geometry oppose the general dogma of the impossibility of motion and warrant that geometric figures can be constructed with a compass and a straight edge, and the axioms rule out Zeno's paradoxes with respect to equality. This reconstruction is plausible if we admit that intuition had a diminished role compared to earlier periods from the beginning, and not just after the discovery of incommensurability. For how would the discovery have come about if intuition had not already been left behind at least temporarily? And how would it have been taken seriously unless intuition was no longer regarded as absolutely reliable? Even a construction such as that in figure 39, which according to Kurt von Fritz shows the absence of a common measure for the sides and diagonals of a regular pentagon, is compelling only if we continue the train of thoughts beyond what is given in intuition (in quantum theory the situation would become indeterminate after a finite number of steps).

Thus, independently of all arguments a new approach to intuition was prepared, one that gave Parmenides' proofs the required momentum. And we saw how fundamentally these arguments went on to alter our thought about nature and humanity.[86]

[28] *Let me sum up*: the Pre-Socratic philosophy of nature introduced into this world a homogeneous substance and a homogeneous plan of construction. Both were further developments of previous ideas, yet are distinguished from these by their greater universality and consistency. They allowed for qualitatively diverse domains, though in such a way that the basic substance was active and the plan of construction valid in all of them. There were no "gaps" where all of a sudden everything behaves differently. The new cosmologies established in this manner did not harmonize with contemporary world intuition; they did not reflect what a contemporary human being saw and experienced in the world. For *Xenophanes* this discrepancy was a sign of the unreliability of all intuition. He demonstrated this lack of reliability using another powerful argument: the god that he had constructed with the help of rules of the understanding that he

---

[86] [*Editors' note: This is where the typescript from the Feyerabend Collection of the Constance Archive ends. The remaining portions of the text are from the somewhat later version that Helmut Spinner made available to us.*]

155

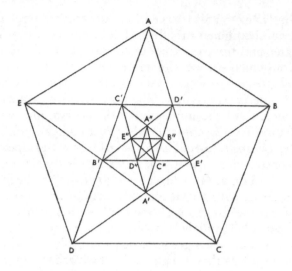

39. Discovery of incommensurability

From: Fritz, Kurt von (1971): *Grundprobleme der Geschichte der antiken Wissenschaft*, Berlin, New York, p. 566, fig. 2.

had just discovered does not have anything in common with the gods whom our intuition pretends to be perceiving in the world. Thus, our intuition has to be cleansed of partiality and superstition, just as ship and bridge construction had been. In this way *experience* became a new, hitherto-unknown, and very strange source of knowledge. Experience has not been a constant measure of our ideas, as naïve naturalism assumes, but appeared relatively late in the history of culture. It is, furthermore, not *discovered* but rather *constructed* on the basis of rules that have nothing to do with the structure of our world. What determined the contents of the new ideologies was not the natural relation between humans and their environment, nor the impressions derived from interactions between the species "human" and nature, directly reflecting the co-existence of humans and their environment, but rather a few abstract rules of the understanding, which – if isolated from the context of language and customs – confronted the natural and social world as something entirely *foreign*. Not only was the new concept of experience foreign, but so was Xenophanes' world picture, which was based on that concept and from which gods and animated principles had been expunged with contemptuous disregard for the consequences.

Parmenides proceeded differently yet with exactly the same result.

He posed his question about the basic substance in the most general way, and he also answered it in the most general way (the most general statement one can make about a thing is that it IS). He then drew his consequences from *this* answer on the basis of very general and abstract rules. Intuition lost its authority everywhere, and humans became increasingly remote from their surrounding nature; even fellow humans were viewed "rationally" and their natural impressions replaced by constructs of the understanding. From then on, conceptual considerations were at the center of studies on human nature and determined the essence of things surrounding us. This is how Western philosophy of nature started. There were no further major transformations, and we can restrict ourselves to an outline of the development up to the present time.[87]

---

[87] [*Editors' note: Spinner remarked on this in the typescript "See Feyerabend letter of 8/26/1974." Unfortunately, this letter is not available, but it appears that Feyerabend had at that time given up on, or at least postponed to an undetermined future time, his original plan to elaborate on the further development of philosophy of nature not in the form of an outline but in two subsequent volumes.*]

WESTERN PHILOSOPHY OF NATURE
FROM ARISTOTLE TO BOHR

The first and immediate consequence of this beginning was the emergence of a multitude of cosmological systems, which can be roughly split into two groups. First, there were those systems that aimed at avoiding the consequences of Parmenides' arguments – there is no motion, there are no parts – with the help of *abstract considerations* while establishing a connection with *specific facts* of intuition (or experience) only later, with great difficulty, and often in a quite artificial manner. In most cases we have to be content with a summary account and vague promises. Intuition is *directly* consulted only at one time, namely when proving the existence of motion and parts. Once this has been proven, intuition is replaced by general observations. Modern science is closely related to this way of thinking.

Second, we have *Aristotle*'s ambitious attempt to build a cosmology that consistently adheres closely to intuition, does justice to Parmenides' arguments, and still is rich enough to incorporate the numerous new facts in astronomy, physics, biology, physiology, mathematics, politics, sociology, art, and history of ideas which had been discovered by Aristotle's predecessors and contemporaries as well as by Aristotle himself. His philosophy has influenced the history of ideas to this day. Aristotelian ideas guided debates in the sixteenth and seventeenth centuries at the beginning of modern science. He was often criticized at the time, but there can't be progress without criticism of such older yet well-worked-out ideas. Aristotelian ideas permeate Thomistic philosophy, which is the foundation of Catholicism and is experiencing a resurgence in contemporary philosophy of nature. Aristotelian ideas are also present in medicine, psychoanalysis, and art, and in the last of these to such an extent that Bertolt Brecht considered it necessary to challenge the Aristotelian theory of drama with his own, non-Aristotelian theory (Brecht 1949b). And in the history of ideas we still have hardly anything better than the brief and concise principles that Aristotle *formulated* in various places in his body of work (*Metaphysics* 993a30ff., 1074b1 ff., *Topics* 183a37ff., *Soul* 403b20ff.) and *used* in his examinations: his theoretical writings almost always start with a report on his predecessors' achievements, not only in order to make the accomplishments of his own philosophy stand out more, but also to drive the argument itself further along. It is not easy to present the wealth, simplicity, and fecundity of Aristotelian philosophy in just a brief outline. Thus, I shall restrict myself to his *theory of motion* and his *epistemology*, and even these I shall not describe in the form in which they appear in Aristotle's writings, where they are often intermingled with both older and more recent material, but rather in an idealized form, which focuses on the *possibilities* inherent in Aristotle's thought. Thus, I shall describe a part of what Lakatos would have called Aristotle's *research program*. From this point onward "Aristotle" will refer to this component of the research program rather than to the historical person (this also goes for all other authors discussed in this chapter).

6.1. Aristotle's Research Program

Aristotle's *theory of motion* included (1) a theory of objects; (2) a definition of motion; (3) a number of laws of motion. The theory, definition, and laws are ridiculously simple if we look at them in isolation. Yet they solved some deep problems and had far-reaching consequences for our knowledge of nature. I shall discuss these points one by one.

(1)   Every *thing* consists of form and matter. This conforms both to our ordinary way of speaking, in which we *ascribe* or – in the case of production – *assign* properties to things, and to appearance: we see ourselves as surrounded by things carrying properties. Form and matter are real qualities that may change but can never exist without each other (there are red lips but there is no redness by itself nor lips without color).

(2)   Motion is "the fulfillment of what exists potentially, insofar as it exists potentially" (*Physics* 201a10). Each motion is an *interaction* between physical systems that takes place in a medium (if no contact is established). It is initiated by a moving force – that is, by an object that possesses an active form – and consists in the passing of this active form onto the moved object, which must be able to receive,

159

to "fulfill," it. A form is active if it has an independent propensity to pass onto a nearby body that does not possess but is able to receive it. Gravity is not an active form because it cannot be passed on to another body capable of receiving it either through contact or via a medium. Heat, however, is an active form. In addition to a mover, motion requires an active form within that mover, a receiver, and a medium; it requires also a gradient between mover and receiver. That is to say, the mover has to possess the form ready to be passed on to the receiver to a greater extent than the receiver. Yet the difference should not be too great, either, lest a normal motion be unable to occur and objects will be destroyed instead. A motion comes to an end once the receiver has the same form as the mover or, to express this process in terms of other Aristotelian notions, as soon as the effective cause (the active form initiating the motion) coincides with the final cause (the final state of the receiver). Thus, every motion in the world has a determinate beginning, middle, and end. We can say that it aspires to reach the end, just as an organism that is out of balance goes through various motions until it has regained its normal condition. The sky's rotation is the sole exception, as it continues infinitely, thereby perpetuating the motions that are in the lower layers of the world.

(3) A consequence of the explanation just given is that motion is not possible without a mover: a body's natural state is rest (which in this case includes qualitative stability). This is the Aristotelian law of inertia. The speed of a change is proportional to the imposed force and inversely proportional to the body's resistance. This is the Aristotelian law of forces.

[30] The following remarks are relevant here. First, this doctrine actually solved the problem posed by Parmenides. The dualism of form and matter makes motion *possible*. Second, it solved the problem without invoking abstract entities that are inaccessible to observation. Both the laws and the concepts used in them were intuitive in the dual sense that they were adopted from ordinary language and could be explained in terms of our ways of intuiting the world at the time. This intuition could also be used to test them. This cannot be said of the non-existent form of being (space) or of the atomists' elements (atoms). Third, the doctrine is *prolific*. The notion that motion does not occur without an external mover is still in use as a research principle today. It was responsible for the scientific discovery of bacteria and other organisms that are not visible to the eye but cause visible changes in visible material. It was also responsible for the Inquisition's notion of an implicit pact between the devil and

witches, that is, of the latter's tacit acceptance of demonic assistance without endorsement by the church.[1] Once it was established that herbs, herbal extracts, minerals, and powders made from them have certain very specific effects, the occurrence of deviating effects (such as curing illnesses by means of poisons) had to be explained by external factors. Where these factors were not directly visible, demonic assistance, and thus an implicit pact with the devil, was suspected.

Fourth, the Aristotelian law of inertia played an important part in the debates concerning the motion of the Earth, which prepared the transition to modern astronomy. None of Copernicus' supporters, including Galileo, had offered a satisfactory answer to the objection that a rock falling from a moving tower would have to hit the ground far from the foot of the tower (*Heavens* 296b22). All of them had *circumvented* the objection by means of verbal adaptations and *ad hoc* hypotheses. Even those medieval thinkers who tried to explain the motion of a rock that has already left our hand, and thus should be expected to fall to the ground, by means of an *impetus*, a moving force imprinted on the rock, still had to rely on the notion of absolute space in which momentum and *impetus* have a precise correlation. Newton was the first to develop a theory of motion able to provide a satisfactory answer to every aspect of Aristotle's argument; in retrospect, he thereby demonstrated the power that this argument had possessed throughout the centuries.

Fifth, the Aristotelian theory of motion gave rise to a theory of elements (*Generation* 329b7ff.) that provided predictions capable of being checked against familiar facts: motion is triggered by bodies that either contain active forms themselves or are composed of bodies containing such forms. Thus from a purely theoretical point of view an element can be defined as a non-composite mover. The number of elements depends on the number of active forms. Aristotle identified four such forms – hot, cold, dry, and moist – and arranged them into the pairs hot–cold and dry–moist (so that motion could occur only if there was a gradient between the elements in the pairs). Intuition tells us that every body has some degree of heat as well as some degree of moisture. Thus (theoretical prediction), there are the following four elements: hot/dry, hot/moist, cold/moist, and cold/dry, and these (empirical confirmation) correspond to the already well-known elements fire, air, water, and earth. This procedure is comparable to the theoretical prediction of empirically discovered elements in the twentieth century. The most astonishing and (for our purpose)

---

[1] Thomas Aquinas (*Summa* II-II, question 96).

interesting results, however, derive from applying the doctrine of motion to the process of perception and knowledge itself.

[31] Modern science relies on experience and experiments to support its laws and theories. This principle gets endlessly reiterated in scientific treatises, commemorative speeches, and polemical remarks, and modern philosophy of science has explored all of its conceivable consequences. Newton had already emphasized that a theory has to provide an accurate account of the phenomena. Bohr, Heisenberg, and their successors in the twentieth century essentially said the same in almost the same words. Yet they failed to clarify *which* processes should be considered phenomena or experiences, nor did we hear anything from them about why phenomena, experiences, and experiments should be given such a crucial role in science. *Modern science has not contributed to our understanding of its foundation or to an explanation of its function.* (This may be because to this day modern science has not found a satisfactory solution to the mind–body problem.)

The situation is different in Aristotle. For him, perception, which has a crucial function in the process of knowledge acquisition, is the result of an interaction that is subject to the general laws of his theory of motion. There is a mover (the perceived object), a medium (air for sounds, ether for light), and a receiving system (the sense organ, which has the ability to take on certain forms of the object such as its color). Once the interaction is completed the sense organ possesses *exactly the same* form that originally triggered the perceptual process, "that which sees [. . .] must be coloured" (*Soul* 425b24); our perception of the environment is temporarily in the grip of the forms that constitute the environment. This temporary subjection to those forms is a consequence of the laws of motion, the structure of our sense organs, and the clarity of the medium between observer and observed. And due to the special nature of our cognitive capacity it turns into *permanent possession*, that is, into *practical knowledge*, after several repetitions of this process:

> So out of sense-perception comes to be what we call memory, and out of frequently repeated memories of the same thing develops experience; for a number of memories constitute a single experience. From experience again – i.e. from the universal now stabilized in its entirety within the soul, the one beside the many which is a single identity within them all – originate the skill of the craftsman and the knowledge of the man of science, skill in the sphere of coming to be and science in the sphere of being. We conclude that these states of knowledge are neither innate in a determinate form, nor developed from other higher states of

knowledge [such as intuitive grasping of isolated ideas], but from sense-perception. It is like a rout in battle stopped by first one man making a stand and then another, until the original formation has been restored. The soul is so constituted as to be capable of this process.

(*Posterior* 100a4ff.; see also *Metaphysics* 980b25ff.)

In the same manner, perception – that is, the physical process just outlined – causes the development of general concepts and thus of knowledge. I observed that the forms received by the sense organ correspond to the forms in our environment only if the medium between the object and the sense organ is able to transport the forms to the observer without disturbance and if the sense organ receives the forms without distortion. Aristotle knew that these two requirements are not always met and that therefore perception is not always reliable. However, he assumed that disturbances were spatially and temporally restricted processes and that perception *overall* provides an accurate reflection of reality. On the whole, the cognizing subject is in harmony with the surrounding world, and its perceptions are free of comprehensive illusions that would otherwise turn the actual order in the world upside down. (This assumption, which covers both the sense organs and the processes between observer and observed object, is what distinguishes Aristotle's cosmology from the cosmology of modern science.) Thus, Aristotle is content with an *enumeration* and *explanation* of possible disturbances: his epistemology is a naïve realism crucially restricted for individual cases (though his followers do not always adhere to these restrictions).

Aristotle was not deluded about the fallibility of observation. According to him, only certain kinds of observations, not all of them, are reliable, as he pointed out in *On the Heavens* 306a18, which according to Owen is to be translated as "perceptual phenomenon that is reliable when it occurs" (*Prior* 46a20).* In *Metaphysics* 1010b14ff. he ranks perceptions on the basis of their reliability and notes the dependency of sense qualities on the state of the perceiving body (*Metaphysics* 1010b21). Even the most basic sensations, which are minimally susceptible to error (*Soul* 428b18), may report erroneously about their own objects because "mistakes are possible in the operations of nature" (*Physics* 199a35) or due to "monstrosities" (*Physics* 199b4). In addition, their reliability depends on the distance of the object from the observer as well as on the extent to which the

---

* [*Translator's note: Here Feyerabend appears to be referring to G. E. L. Owen's influential paper "Tithenai ta Phainomena," which was first published in S. Mansion (ed.), Aristote et les problèmes de méthode, Louvaine, 1961 (p. 90 n. 13).*]

observer is familiar with the object (*Parts* 644b25). Notably, the sense organs can also be triggered by inner stimuli, so that the resulting fabricated sensations have no relation to the external world (*Sleep* 460b23). In normal perception, however, the sense organs do not merely receive the arriving forms passively; rather, they also react to them (*Sleep* 460b25). Aristotle occasionally rejected even clear and irrefutable impressions such as dream images due to their improbable causal chains (*Prophesying* 462b14). In doing so he admitted that *even false beliefs can be supported by experience*. Conversely, he also drew our attention to natural processes that are imperceptible due to the minuscule size of their objects (*Meteorology* 355b20).

Thus, we see that Aristotle was more familiar with problems of perception and of knowledge based on it than many a "modern" philosopher. Yet he refused to prioritize speculative facts over basic facts of perception. He did not consider "forcing [our] observations and trying to accommodate them to certain theories and opinions of [our] own" advisable (*Heavens* 293a27). Nor did he think it was a good method to be "led to transcend sense-perception, and to disregard it on the ground that 'one ought to follow the argument'" (*Generation* 325a13). With regard to *basic issues* such as questions concerning motion (Does motion exist? Is everything always at rest?), perception, once cleared of the aforementioned minor errors, *always* has the last word. This is what distinguishes Aristotle from thinkers such as Kepler or Galileo, or at least from the ways those thinkers put their theories into practice (the empiricist phraseology was upheld by both, after all). In any case, Aristotle's epistemology was not a separate discipline but a *natural consequence of his general principles of philosophy of nature*; it tells us which processes in nature qualify as experience and why we can rely on them.

[32] Despite the simplicity of its basic principles Aristotle's philosophy of nature is astonishingly successful in dealing with complex physical and mathematical problems. One example, which will not be further discussed here, is the problem of the *continuum* (*Physics* 231a23ff.) and the related solution to Zeno's paradoxes (*Physics* 230b28ff.). Another example is the Aristotelian *theory of light*. According to Aristotle the object of seeing is that which is visible, i.e. *color*. Color

> sets in movement not the sense organ but what is transparent, [. . .] and that, extending continuously from the object to the organ, sets the latter in movement. Democritus misrepresents the facts when he expresses the opinion that if the interspace were empty one could distinctly see an ant on the vault of the sky; that is an impossibility. Seeing is due to an

affection or change of what has the perceptive faculty [since the eye is not in the location of the object itself], and it cannot be affected by the seen colour [which is located in the object] itself; it remains that it must be affected by what comes between. Hence it is indispensable that there be something in between.

(*Soul* 419a13)

The medium moves only when it is in a state of readiness to move, or when it is "transparent." It becomes transparent under the influence of fire (*Soul* 419a24; see also *Sense* 439a20). And it is this *state* of transparency of the medium, rather than a movement *caused* by the medium, that Aristotle identifies as light (or better, as brightness): "light [brightness] is the activity – the activity of what is transparent so far forth as it has in it the determinate power of becoming transparent; where this power is present, there is also the potentiality of the contrary, viz. darkness" (*Soul* 418b10). If the medium loses its transparency (its "color," as Aristotle puts it in *On Sense and the Sensible*) it thereby also loses its ability to be affected by the colors of the objects, and then these colors and with them the objects themselves become invisible. Darkness sets in. The medium needs the quality of brightness in order to pass on the motions caused by qualities in objects such as their colors. It is not easy to assign a specific speed to the quality of brightness thus explained. It would contradict the observable facts and also "reason" (*Soul* 418b24) – or as we would say today, it would contradict the grammar of the word "brightness."

Overall, this theory accurately describes our everyday experience: colors are in the object, and we see them only when there is light. Light is not in any specific object, for if it were then only that particular object would be visible, nothing else. However, all objects in an extended area are visible, hence light is associated with that area. Furthermore, it is not an extended substance, for such a substance would have its own peculiar qualities, and that would diminish visibility rather than facilitate it. But if brightness (or light) is neither a body nor an extended substance then it can only be a quality, and this is indeed Aristotle's conclusion: brightness is the quality of being transparent, and this quality is actualized in the medium in the presence of fire.

Aristotle's theory prohibited the formulation of sentences that pose no problems for us today. Speaking of "red light" made no sense for Aristotle. For him redness was a quality of objects, and light was not a body but a quality of the space between objects. From that perspective it is not easy to figure out what "velocity of light" is supposed to be. Colors do not have a velocity, for they are located in bodies.

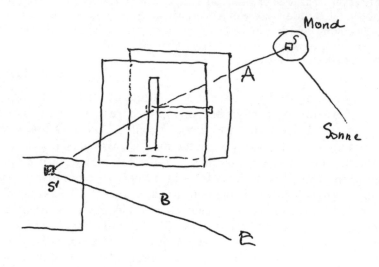

40. Alhazen's studies on light I

Drawing by Feyerabend.

However, we have two options in speaking of a velocity within the medium. On the one hand, we have the expansion of the light (assuming that such a process of expansion takes time), and on the other hand, we have the motion of the disturbances going on *within* the light. These *conceptual problems* that come up when we try to assign velocity to light within the Aristotelian theory cannot be solved with the help of experiments, and it is extremely naïve to ascribe such an experimental solution to Galileo.[2] Galileo's experiments are entirely unintelligible without a detailed conceptual discussion. This explains why certain studies of the phenomenon of light today appear to us so awkward and trivial.

One example is Alhazen's studies as reported by Schramm (1963: 229ff., 80ff.). In his study of moonlight Alhazen showed that each point on the moon's surface perceived by the eye in direct vision can also illuminate a screen linked to it by a straight line. This eliminated the hypothesis that the moon's surface was a purely subjective appearance such as an extended glare episode.[3] Notably, the Aristotelian

[2] As Cohen did in section 7 of his paper "Roemer and the First Determination of the Velocity of Light" (1940: 331ff.).
[3] Plutarch, *Face*, p. 17 in the Cherniss translation.

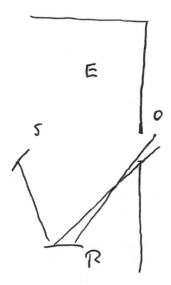

41. Alhazen's studies on light II

Drawing by Feyerabend.

theory (which has nothing to do with the aforementioned hypothesis) has to be either expanded or modified.

The screen is visible only in those locations S' that can be connected by a straight line to a point S in the moon. The light point S' disappears as soon as an obstacle is placed in the way A. If the eye is located in S' then the Aristotelian theory is applicable without any difficulties. If the eye is located in E then we can only say that the medium must be transparent along B. And yet the path A still has to be kept open if we wish to see the spot S' from E. Alhazen at first follows Aristotle and distinguishes lines of *visibility* such as S'E, where the eye is in E, or S'S, where the eye is in S, from lines of *radiation* such as SS', where the eye is in S, and lines of radiation such as SS', where the eye is in E. But he soon considers the hypothesis that processes take place on both lines, namely processes of the same kind. The following arrangement [figure 41] reflects the problem even more pointedly.

Sunlight enters a dark room through an opening O and illuminates a red carpet R. The eye E sees the red carpet because the space between R and E has become transparent. Yet a white screen S, which unlike the sense organ does *not* have the capacity to become red, also appears red, namely in all those locations that can be connected with the red

167

carpet by a straight line. Let 1 be one of these lines. Then according to the hypothesis just formulated a certain process has to take place between S and R. Now S appears red. Hence the process has to be sufficiently specific to transmit red. Thus, we are forced to assume that colors move, and this assumption creates problems for the Aristotelian theory. Alhazen drew the following conclusion: (a) the medium along the lines emanating from a luminous object is transformed, (b) the transformation is more specific than a mere transition to transparency, and (c) it can be usually detected only by the eye – hence the initial plausibility of Aristotle's theory – but (d) strong radiation can also illuminate screens and in general any kind of object. This conclusion was reached indirectly via many detours, and the detours show us just what a huge *conceptual* distance separates us today from Aristotle's theory.

[33] The rise of modern science led to a new approach to nature and to the status of humans within nature. More precisely, we have not *one* new approach but many; it was not *one* argument or *one* group of facts that led us into the seventeenth century, but an entire arsenal of reasons, facts, prejudices, social pressures, samples of intuition, and problems in physics, astronomy, and theology. The development passed through a stage – early Christianity – in which a closer approximation to the mythological phase occurred, since the conceptually explicit was considerably reduced in this stage. Yet soon heresies forced us to develop clearer formulations of basic truths, delineations, and distinctions, thus initiating a new phase of conceptuality which focused initially on the Christian mysteries and subsequently on the entire domain of secular knowledge. And slowly we moved toward Aristotle again, though toward a version of Aristotle that was more schematic and dogmatic in presentation and understanding. Nineteenth- and twentieth-century research has shown how complex the processes that led from there to modernity are, and to what extent the development was once again driven by non-conceptual factors.

An example of such factors is the work of the brothers Limburg in the early fifteenth century, in which events in salvation history such as Christ's crucifixion are not isolated from their secular context but subject to the same laws. "It was a pictorial anticipation of Giordano Bruno's pantheism. An important insight is [. . .] that a reorientation of the human mind is often expressed in the domain of fine arts before it can achieve conceptual clarity in the domain of philosophy" (Roth 1945: 57). We do not yet have a conclusive assessment of the efforts in social history, conceptual history, history of religion, history of science, and "general" history (that is, political history and history of warfare) to understand this transition. And yet even today we can already see that most existing assumptions about the rise of modern science are not only historically false but at times even logically impossible. The body of material is too extensive even for a sketchy outline. Therefore, I will only be able to *highlight* aspects of this development without claiming any completeness, and also without any guarantee that the episodes, ideas, theories, and cosmologies discussed really played a significant role.

The highlights will illuminate, first, the *mathematical approach to nature*; second, the *absence of a foundation*; and third, the constant *dynamics of concepts*, which eventually becomes part of the dynamics of the physical systems themselves. I shall illustrate the first aspect through the example of Descartes, the second through the examples of Bacon and Galileo, and the third through the examples of Galileo and Hegel. Hegel presented a comprehensive theory of the new role of concepts (while at the same time having a very imperfect grasp of their mathematical side). He provided a transition to the dialectic approach to nature, which was developed *theoretically* by Engels in the nineteenth century and *practically* by Bohr and Einstein. In what follows, philosophy of nature (science) will be discussed in close conjunction with epistemology.

6.2. Descartes: The Mathematical Approach to Nature

[34] Mathematical considerations are quite common in Aristotle's work, as for example in his explanation of rainbows (*Meteorology* 373a32ff.). But such considerations always have only an instrumental function. They allow us to predict the behavior of things whose *nature* has been already determined in other ways. This separation of basic observations and instrumental predictions, which is already implicit in Aristotle, was later codified by Simplicius:[4] *physics* describes the nature, position, size, and shape of objects; it covers the principles of motion and provides reasons for motions that occur. *Mathematics* (*astronomy* in Simplicius' context of scientific problems), by contrast, is content with the prediction of phenomena whose nature does not have to be more fully known, and it can complete its tasks without any such knowledge. Furthermore, the requirements of a discussion in physics include that the concepts used always be close to empirical intuition (to ordinary language) and that the principles be based on experience.

---

[4] Simplicius, *Physics*: 291ff.; see Duhem's masterly essay (1908).

At first, *Descartes* also imposed certain restrictive conditions on his physics. He, too, required that both the basic concepts of physics and the acceptable methods used in this science should be close to intuition. And yet the kind of intuition that he selected as his criterion of restriction was very different from Aristotle's intuition. First, it was of a mathematical nature (internal intuition), and second, it is not immediately accessible but requires a course of self-discipline in order to be *revealed*. This is the function of the skeptical considerations in the First Meditation, which readers not only have to understand but must also allow to affect them emotionally. And therein lies a significant difference between Descartes' physics and Aristotle's. Only mathematics (geometry) meets the requirements, hence physics has to be built entirely of geometric concepts. Descartes guaranteed the *feasibility* of such an edifice by identifying space (the medium of geometry) with matter, and by selecting motion as the distinguishing element.

A cylinder A rotating within volume V consists of exactly the same material as V, namely space material, yet it is objectively (though not always observably) separate from the rest by its motion. The spheres K all consist of the same material as their environment, yet they are all separated by their environment and united into one continuous body. This body affects other bodies not by the density of its substance but by its motions, and this is also how it becomes noticeable. Light rays bounce off it, not because it is less impenetrable than its environment but because it is in motion and because each motion, including the motion of a light ray, is also subject to a principle of conservation. Descartes attempted to construct his universe from space and motion alone, making him more economical than the atomists, who added atoms to the mix. The task was introduced in the *Principles of Philosophy*, but it wasn't really tackled due to the many ancillary assumptions that kept creeping in. Furthermore, Descartes wanted to justify his reduction of cosmology to geometry philosophically – an attempt that failed so badly that according to Leibniz he would "have better omitted" it (Leibniz, *ad Cartesii* II, 1). Nonetheless, Descartes' approach initiated a new phase of the study of nature. The attempt to grasp nature by means of mathematically recordable and empirically testable models started to gain a universal following.[5]

Descartes expressed himself very clearly on this matter. The study of nature can be accomplished in two different ways, he wrote. We can *explain* a particular process in nature by reducing it to first principles – that is, we can show what motions of space material produce it – or

---

[5] Crombie (1953) contains a discussion of the predecessors.

42. Descartes' cylinder
Drawing by Feyerabend.

we can put forward hypotheses which, though in accord with the first principles, have not been derived from them and may not be derivable from them at all due to the applicability of special principles, but which nevertheless describe the process in an empirically adequate manner. Descartes chose to put forward hypotheses so as to not appear "too presumptuous" (Descartes, *Principles* III, 44), or as illustrations to explain the manner in which the basic principles work in special cases (Descartes, *Discourse* V; *Principles* III, 45). He occasionally introduced a false hypothesis deliberately if in his view such a move would lead to a better understanding of things than the truth:

> [T]aking into account the omnipotence of God, we must believe that everything he created was perfect [partly meaning "complete," "finished"] in every way. But, nevertheless, just as for an understanding of the nature of plants or men it is better by far to consider how they can gradually grow from seeds than how they were created [entire] by God in the very beginning of the world; so, if we can devise some principles which are very simple and easy to know and by which we can demonstrate that the stars and the Earth, and indeed everything which we perceive in this visible world, could have sprung forth as if from certain seeds (even though we know that things did not happen that way); we shall in that way explain their nature much better than if we were merely to describe them as they are now (or as we believe them to have been created).
>
> (*Principles* III, 45)

171

Occasionally, as in the case of light, Descartes even offers us both a fundamental explanation (*Principles* III, 55, 68) – light is a pressure spreading evenly and instantaneously in all directions from the center of a vortex, whereby this pressure exists even if there is "no [light-producing] force in the stars themselves" (*Principles* III, 64) – *and* "two or three comparisons" (Descartes, *Optics* I). The comparisons show that even *perception* in Descartes and his successors is a very different process from Aristotle's perception. It is not one of the identical *forms* active within the object itself that enters the sense organ, but a *motion*, which possesses just as many degrees of freedom as there are distinguishable perceptual qualities without resembling these qualities in any way. The same impact that causes "an infinity of fireworks and lightning flashes" in the eye causes a sound in the ear and pain in other parts of the body (Descartes, *Optics* VI). It is not even necessary for an object or substance to move toward the eye during the process of transmission. A blind person surveying her environment with a stick is able to determine its structure by means of the stick's *motions* without receiving a piece of matter or a form from the environment itself. But if the effect, perception, is so fundamentally different from its cause then the former cannot be used to draw conclusions about the qualities of the latter. This is why the fundamental principles have to be derived from a source other than perception (*Principles* IV, 203).

It is interesting that Descartes' arguments are based on facts that were familiar to Aristotle as well and that the difference between the two thinkers arises solely from their different *interpretations* of these facts. From the fact that a process causes sparks in one's eyes, a sound in one's ears, and pain on one's skin, Descartes concludes that the senses *never* accurately reflect their environment and that even in the standard case the perceived object is fundamentally different from the object causing the perception. For Aristotle, by contrast, those examples represent cases in which nothing is perceived in the first place, and they do not support any conclusions about the processes that take place in normal perception. We might then be tempted to give priority to the Cartesian theory due to its greater *generality*. But that would also be a mistake. For Aristotle does not assume that the eye randomly reacts sometimes in one way and sometimes in another. Rather, he assumes that in a standard case the forms are transmitted *almost* free of distortion, and that the characteristics of the eye as the medium, which after a physical blow become noticeable in isolation without any cognitive function, *are effective also in a standard case* of perception, resulting in minor, innocuous disturbances. At the same time, Descartes lacked a theory of interaction between mind and body

172

that would explain *why* the particular qualities that we perceive in our conscious mind are triggered by the external disturbances. Here is a gap that even more recent research has not been able to close. There is no such gap in Aristotle's system; the perceiving component of the mind is subject to the same laws as the physical bodies in the world, and both take on the same forms.

If we accept the Cartesian program then we obtain a tool with which to understand nature that is simple in principle yet highly powerful in its application. The problem is not the *invention* of models but the selection of a suitable model from among a multitude that our faculty of imagination keeps ready for every conceivable process.

> I freely venture to state that I have never observed any [objects presented to my senses] which I could not satisfactorily explain by the principles I had discovered. But it is necessary also to confess that the power of nature is so ample and vast, and these principles so simple and general, that I have hardly observed a single particular effect which I cannot at once recognize as capable of being deduced in many different modes from the principles, and that my greatest difficulty usually is to discover in which of these modes the effect is dependent upon them; for out of this difficulty I cannot otherwise extricate myself than by again seeking certain experiments.
>
> *(Discourse VI, 6)*

Compare the nineteenth-century situation in which different models were used to explain thermal and electricity phenomena. Until Einstein, science and a large part of natural philosophy retained this scheme to explain natural phenomena because of its simple elements and their interactions. The elements changed and the laws of interaction changed, especially under Newton's influence, but the idea that objectively – and especially mathematically – describable mechanical models (with or without fields) sufficed to explain physical, biological, and psychological phenomena remained.

6.3. Galileo, Bacon, Agrippa: Empiricism without Foundations

[35] The second characteristic trait of the modern way of studying nature is the absence of a *foundation*. Though there is a lot of *talk* about the new and fertile foundation that Descartes, Galileo, and Newton introduced and used in their research, such a foundation cannot be found in *practice*. *All* concepts change, including observational concepts as well as clear and distinct ideas of internal (mathematical) intuition. Unlike Aristotelian philosophy, which *requires, identifies,*

and uses a foundation *in actual research*, we now have a fundamentalist epistemology and an entirely separate practice of research in philosophy of nature and science. This antagonism and the related irrationality of modern science is hidden by a slanted method of representation, which depicts even the most revolutionary discovery as resting on a solid foundation. Often the only difference between science and philosophy of nature is this: one makes discoveries, and the other one relates these discoveries to a foundation. Let me briefly outline two episodes of Galileo's research as an example of this new practice.

According to the Aristotelian tower argument, a rock falling down the side of a tower that is moving with the Earth will hit the ground far away from that tower's base. This is a direct consequence of the principles of Aristotle's theory of motion discussed in section [29] [*chapter 6.1*]. Experience does not show us any such occurrence, hence the Earth is at rest. The tower argument had a significant function in the Middle Ages as well as in discussions of the Copernican doctrine. Copernicus mentioned it in chapter VIII of his major work *On the Revolutions of the Heavenly Spheres* (1543). Galileo dispelled the force of the argument in two steps. *First step*: from his predecessors he adopted the impetus theory according to which an external mover transmits an internal mover, a so-called impetus, to the object moved by it. This impetus continues to move the object in the original direction even after contact with the external mover has been terminated. *Second step*: he separated motion and activity and

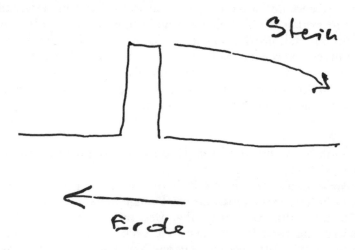

43. The tower experiment

Drawing by Feyerabend.

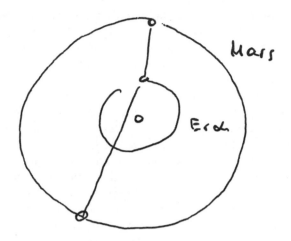

44. The brightness of Mars

Drawing by Feyerabend.

turned the former into a *relation* between an object and a frame of reference. This second step moved far away from sensual perception, according to which the ship but not its port moves, since a moving force is active only in the ship (Aristotle, *Sleep* 460b25). Subsequently even observation statements about motion had this new, abstract meaning – a clear change in the observational concepts common in the seventeenth century, which Galileo concealed with his tendentious presentation and his frequent recourse to the Platonic theory of anamnesis. Note also that this change did not take place in the "theoretical superstructure" but in the "empirical base." Crucial concepts within this base were replaced for the purpose of promoting the inner harmony of the new worldview. Any change promoting this harmony was permitted, even a change in the sensory component of our perception.

This leads us to the second episode in Galileo's attempt to isolate the Earth's motion from the Aristotelian context and to integrate it into a context in which it would be possible: in Copernicus the distance between Mars and Earth changed considerably. This change in distance was not accompanied by a corresponding change in the brightness of Mars, at least not according to observation.

Galileo showed that the brightness of Mars, as modified in the telescope, changes in the desired manner. This sufficed for him to give priority to the telescope over the naked eye despite the absence of a

175

theory of telescopic vision. Once again an element of observation – this time an element of sensation rather than a concept – was changed in order to ensure the survival of an attractive idea. And all of this happened while we were verbally still clinging to the idea of an observational basis for our knowledge.

[36] Galileo developed the epistemology underlying this seemingly rather arbitrary proceeding only in outline; Bacon presented it more systematically. *Bacon* was the first modern philosopher who at least took into consideration the notion of research without a foundation. He trusted neither experience nor thought. Both were in need of reform and transformation into clear, distortion-free mirrors of the physical world. A reformation of experience and thought, however, was not possible without a reformation of humanity itself, since experience contains *natural* human reactions to the environment, and thought *natural* responses to such reactions. Thus, the human being itself had to be remade; a new human being had to be introduced with new sense organs and new ideas that would turn it into a harmonious part of nature rather than separating it from nature by prejudice and distortive sensations. The transformation begins with the removal of inaccurate basic principles and proceeds by gradually establishing new ones. Thus, *during* the reform process (which can take centuries) no foundation is available, and Bacon believed he had discovered a method to connect us directly with reality, without the mediation of perceptions, sensations, or ideas. He was thinking of the experimental method and anticipating a state of science which largely excludes experience from the process of data collection concerning natural phenomena: not experience, not perception, but a complex physical system, along with experimental arrangement, investigates the properties of elementary particles; experience is used only to determine the *result* of the investigation and to deliver it to the human mind (Feyerabend 1969b). In this respect Bacon's ideas were truly groundbreaking! Let us, then, take a closer look at them!

Bacon's reform of knowledge, and with that also of views about nature, proceeded in two steps, of which one was negative and the other constructive. The negative step had a lot in common with contemporary foundationalism in religion. Just like the Protestants, Bacon believed that the rebuilding of knowledge had to be preceded by a work of "instauration" (Bacon 1620: §115), by "expiations" or "purgings of the mind" (§69), which were to "sweep away" (§97) the familiar proofs and theories, and which would also remove all those generally known and used concepts "which follow [. . .] the act of sense" (Bacon 1620: Preface) so closely that the senses themselves seem to be speaking through them. The entire "work of the mind" is restarted from scratch, but only after "the floor of the mind" has been "swept and leveled" (§115), so that a truthful reflection of the processes in nature can be obtained. Preconceived notions (§36), views (§42), even the simplest and most familiar words (§§59, 121) "all [. . .] must be renounced and put away with a fixed and solemn determination, and the understanding thoroughly freed and cleansed; the entrance into the kingdom of man, founded on the sciences, being not much other than the entrance into the kingdom of heaven, whereinto none may enter except as a little child" (§68).

This "destructive" part of the process is in many ways similar to the Protestants' destructive work, which began a century before Bacon. They, too, assumed that our true knowledge of God and His properties had to be cleansed of prejudices and human despotism as well as council decrees, papal dogmatism, and philosophical speculation, then reconstructed by reducing it to the teachings of the Bible. "For that which is asserted without the support of the Scriptures, or of an approved revelation, it is permitted to hold as an opinion, but it is not necessary to believe," wrote Luther (1520: 280). According to Calvin,

> they see manifest signs of God speaking in Scripture. From this it is clear that the teaching of Scripture is from heaven. And a little later we shall see that all the books of Sacred Scripture far surpass all other writings. Yes, if we turn pure eyes and upright senses toward it, the majesty of God will immediately come to view, subdue our bold objection, and compel us to obey.
>
> Yet they who strive to build up firm faith in Scripture through disputation are doing things backwards.
>
> (Calvin 1536: 23)

Compare this with Bacon's assessment of the Ancients:

> The sciences which we possess come for the most part from the Greeks. [. . .] Now the wisdom of the Greeks was professorial and much given to disputations, a kind of wisdom most adverse to the inquisition of truth. [. . .] Assuredly they have that which is characteristic of boys: they are prompt to prattle, but cannot generate; for their wisdom abounds in words but is barren of works.
>
> (Bacon 1620: §71)

Thus, disputations, discussions, and examinations of different opinions obstruct true knowledge and should be eliminated: "we should at once and with one blow set aside all sciences and all authors" (§122). Only in this way can we achieve "the knowledge of simple natures

177

[which] well examined and defined *is as light*: it gives entrance to all the secrets of nature's workshop, and virtually includes and draws after it whole bands and troops of works, and opens to us the sources of the noblest axioms; and yet in itself it is of no great use" (§121, my emphasis). In both cases the object of criticism is intellectual discussion, in both cases we are asked to give it up, and in both cases we are promised "immediate perception," either of God or of nature.

Regarding the *constructive* part of Bacon's philosophy we should note that he did not advise us to start with experience or sensation, as some of his successors believed. He ruled out sensation altogether and replaced it with a method of grasping nature directly without recourse to human reactions: "for the sense by itself is a thing infirm and erring" (§50).

> For it is a false assertion that the sense of man is the measure of things. On the contrary, all perceptions as well of the sense as of the mind are according to the measure of the individual and not according to the measure of the universe. And the human understanding is like a false mirror, which, receiving rays irregularly, distorts and discolors the nature of things by mingling its own nature with it.
>
> (§41)

This is a very different assessment of the senses from that of Aristotle, who thought that external forms reappear in the sense organ almost entirely undistorted. Bacon repeatedly brought up the "dullness, incompetency, and deceptions of the senses" (§§50, 52), reducing sensation to the task of "touching the experiment only" (§50), while it is in turn the experiment that "touch[es] the point in nature and the thing itself" (§50). This completely coheres with modern methodology. Thus, when Bacon spoke of a "well-purged mind" (§97), he did not mean sense data or "raw feels" or any other monstrosities of modern philosophy. Rather, he meant the reactions of a sense organ that was purged of all special theories and characteristic reactions (sensations), and so was able to testify as to the nature of the object rather than the nature of the observer: "well-purged senses" are the senses of the new human of the future who has overcome his own nature and transformed his thinking with the help of experiments in such a way as to finally fit harmoniously into the nature around him. This new human was very different from Aristotelian humans whose natural experience reflected the world more or less accurately, and likewise the world of the new human was quite different from the Aristotelian world of finite qualities and stepwise motion. Thus, the new philosophy of nature transcended natural human capacities, which Aristotle isolated from their mythical context and made explicit, and paved the way for the dominion of *artificiality*.

Bacon believed that he had discovered a method that brings in results even after all theoretical assumptions have been eliminated and in which sensation and ordinary experience do not play a significant part. He believed that this method would enable him to circumvent the existing erroneous human subjectivity and to establish direct contact with nature or, as he calls it, with the "particulars" (§36), which appear to be the same as individual objects in nature. Perception acknowledges the *results* of this method; it records what the method has helped discover. *Yet it is not involved in the process of discovery itself.* Bacon hoped that the method would in the end reform even perception itself, transforming humans from dreamers and debaters into useful observers who represent "a closer and purer league between these two faculties, the experimental and the rational (such as has never yet been made)" (§95). It is clear that this method would not use a "foundation" in the standard sense. The adaptation of humans to nature is a process that leads to knowledge while not resting on any fundamental knowledge itself.

Let me say a few things about the method itself as Bacon understood it. The method was "mechanical" (Bacon 1620: *Preface*); that is, anyone can use it and obtain results. The method was not empirical in the way in which it was understood by some of Bacon's successors. "The empirical school of philosophy," wrote Bacon about the empiricists,

> gives birth to dogmas more deformed and monstrous than the Sophistical or Rational school. For it has its foundations not in the light of common notions [. . .] but in the narrowness and darkness of a few experiments. To those therefore who are daily busied with these experiments and have infected their imagination with them, such a philosophy seems probable and all but certain; to all men else incredible and vain. Of this there is a notable instance in the alchemists and their dogmas.
>
> (§64)

Nor are "endless repetitions" of experiments to be recommended (§85).

> For the induction which proceeds by simple enumeration is childish; its conclusions are precarious and exposed to peril from a contradictory instance; and it generally decides on too small a number of facts, and on those only which are at hand. But the induction which is to be available for the discovery and demonstration of sciences and arts, must analyze nature by proper rejections and exclusions; and then, after a sufficient number of negatives, come to a conclusion on the affirmative instances.
>
> (§105)

179

For Bacon such a "store of particulars" (§103) that pays sufficient attention to negative instances was the starting point of scientific research. It was not supposed to be restricted to a small area, such as astronomy, but was to be created entirely independently of the existing division of scientific disciplines (compare the very similar directives for the collection of Machian elements). The knowledge derived from such a collection would be more comprehensive, rational, and adequate than the knowledge offered by the existing theories. The Copernican system and the Ptolemaic system

> give [...] us the number, situation, motion, and periods of the stars as a beautiful outside of the heavens; whilst the flesh and the entrails are wanting: that is, a well fabricated system; or the physical reasons and foundations for a just theory; that should not only solve phenomena; as almost any ingenious theory may do; but shew the substance, motions, and influences of the heavenly bodies, as they really are.
>
> (Bacon 1605: section IV: *On Physics*)

Thus, here dynamics serves as a criterion of selection among the many possible but relatively uninteresting kinematic constructions for the planetary orbits populating the astronomical publications of the time. Again, experience did not matter.

Bacon's "empiricism" – an empiricism based on a continuously changing notion of experience until harmony between man and world is established – can be fully appreciated only in comparison with other contemporary theories. An interesting version of empiricism, which even influenced the great Newton (though Galileo was not very impressed by it), derived from the occultist tradition – a tradition Bacon radically opposed and which "is to be sharply distinguished from the rational movement that dominated medieval philosophy, theology, and *scientia*" (Schmitt 1969: 86).[6] Heinrich Cornelius Agrippa, Johannes Trithemius, the legendary Faust – they all refer to the fact that reason has its limits and occasionally requires support from a mysterious, magical, and yet reliable source, namely experience. According to Agrippa, formal qualities "are called occult qualities, because their causes are hidden, and because human intellect cannot entirely investigate them – whereby the greater number of philosophers attained this from very long experience, rather than from searching by reason" (Agrippa 1533: I, 10). This magical source is usually combined with certain useful actions to result in discoveries and practical progress: the call for

---

[6] This article also contains additional references on the magical component of empiricism.

experiments rose in the occultist tradition and in alchemy long before the arrival of Bacon's philosophy, which reinterpreted it for very different objectives.

The first part of Agrippa's *Occult Philosophy* covers the elements and their sensible qualities. Agrippa only occasionally had recourse to experience, not because he considered observation as insignificant but because he considered it self-evident that most qualities of the elements are already familiar from everyday observations. "[W]ater is so much necessary," he wrote (Agrippa 1533: I, 6*),

> that without it no animal can live, and no herbs or plants can sprout forth without first moistening with water. It is the seminal virtue of all things: first in animals, whose seed is distinctly watery, but also the seeds of fruit and herbs, although they are more earthy. Yet it's necessary to dissolve [them in] water, if they should be fruitful – or if they absorb the moisture in the earth, or dew, or rain, or water added on purpose.

It is only in connection with occult or hidden qualities that experience features as a *special* source of knowledge:

> [Apart from the virtues just mentioned, which characterize individual elements and are recognizable by means of the senses] there are also other virtues in things that are not from any element, such as driving away poison, driving away ulcers, attracting iron, or other things. These virtues are the result of the appearance and form of this or that thing [as opposed to its constellation of elements and their sensible qualities], from which also being[s] small in quantity are not small in conducting effects, which is not given to any elemental quality. [. . .] They are called occult qualities, because their causes are hidden, and because human intellect cannot entirely investigate them – whereby the greater number of philosophers attained this from very long experience rather than from searching by reason.
>
> (Agrippa 1533: I, 10)

Thus, there are qualities that can be ascertained via the senses and understood through reason. And there are other qualities that, while they can still be investigated by means of the senses, are not accessible to reason. Reason has access to only a part of the world, and far more of it is accessible to sense experience.

---

* [*Translator's note: Feyerabend erroneously refers to ch. 8 here, but the text quoted is from ch. 6. He appears to have translated the Agrippa passages directly from Latin into German, emphasizing that the (Latin equivalent of the) phrase "as everybody can see" occurs frequently in the treatise. It is omitted in the modern English translation of the above passage, which in other respects is much more readable than the first published English translation of 1651. One phrase was replaced here to make sense of the sentence and to better match the meaning of the original Latin wording.*]

It is interesting to compare this view with Galileo's in his early treatise *On Motion*. Galileo's opponent was Aristotle rather than the occultist tradition. Yet the argument is the same in both cases. Galileo frequently appealed to experience to support his own arguments and challenge the views of others that he opposed. But note that on occasion he also clearly delineated the limits of experience and emphasized that philosophers of nature have to use other intellectual techniques as well. This becomes obvious from two passages early on in *On Motion*. The first passage criticizes Aristotle for relying too heavily on experience in his analysis of the motion of bodies through a medium:

> Aristotle wrote (*Physics* 4, 71) that the same body moves more swiftly in a rarer than in a denser medium, and that therefore the cause of slowness of motion is the density of the medium, and the cause of speed is rareness. And he asserted this on the basis of no other reason than experience, viz., that we see a moving body move more swiftly in air than in water. *But it will be easy to prove that this reason is not sufficient.*
>
> (Galileo, *Motion* 24 – my emphasis)

After disproving the Aristotelian position to his own satisfaction Galileo goes on to explain his own position. He begins his exposition with the following words:

> But, to employ reasoning at all times rather than examples (for what we seek are the causes of effects, and these causes are not given to us by experience), we shall set forth our own view, and its confirmation will mean the collapse of Aristotle's view.
>
> (*Motion* 27)

Thus, for Galileo science deals with the "causes of effects," and these are not accessible to *experience* [*experimentia*] – or at least not to experience alone. Galileo appeared to hold the following view regarding this matter: experience is often useful if we wish to end a particular dispute. By observing the world around us we are sometimes able to decide for or against a certain opinion that was presented to us. Thus, we can sometimes criticize Aristotle for holding views that do not match our experience. Then again, Aristotle relied too heavily on experience and did not assign sufficient weight to reasoning [*rationes*], while according to Galileo proofs are conducted by means of *reasoning*. That is to say, demonstrations and proofs are based on "ratio" rather than on "experientia" (Schmitt 1969: 110ff.).

Thus reason and experience switched roles here. For Agrippa

182

reason had its limits, while experience was able to access areas that were beyond reason's reach. Conversely, for Galileo experience had its limits. It was a useful propaganda tool against the Aristotelians, who were all too impressed by the results of observation, but it needed to be supplemented with reasoning to explain how it can be used as well as to arrive at additional causes.

An experience-based philosophy of the simplest kind was the art of witch discovery, which was practiced in Bacon's own time in his own country. It was empirical, it used the enumeration of individual cases as an argument, and it neglected both negative instances and the possibility of misperceptions. Master Matthew Hopkins – the outstanding, wise, and awe-inspiring "Witch-Finder General" of the 1640s – answered the question of whether his ability to find such a large number of witches derived "from his profound learning, or from much reading of learned authors concerning that subject" as follows: "From neither of both, but from experience, which though it be meanly esteemed of, yet the surest and safest way to judge by" (Hopkins 1640: answer to query 3). The method of witch discovery indicated in figure 45 from a Matthew Hopkins pamphlet (swimming in the indicated position) was very popular and was discussed and defended in medical treatises.[7]

This already comes very close to Royal Society empiricism, which found its expression in a museum "into which flowed a strange mixture of objects of real value with others of only passing interest; and the passion for collecting biological freaks brought many worthless and even fraudulent objects into its cases" (More 1962: 500). Among the museum's curiosities (which it shared with many contemporary pharmacies, themselves often small museums on the side) were "an ostrich, whose young were always born alive; an herb which grew in the stomach of a thrush; and the skin of a moor, tanned, with the beard and hair white" (Weld 1848: I, 216). This flea-circus empiricism was also displayed in many experiments that appeared to have no objective other than to reach an inconclusive result with respect to some general superstition. One example of this is the following: "July 24. A circle was made with powder of unicorn's horn, and a spider set in the middle of it, but it immediately ran out several times repeated. The spider once made some stay upon the powder" (Weld 1848: I, 113). Eyewitness reports were trusted almost without challenge, and the narratives of sailors – uneducated, unbiased, and

---

[7] An early discussion is the work of Scribonius (1583) as well as the objections by Hermann Neuwaldt (Neuwaldt, Scribonius, and Julius 1584).

45. The witch swims

From: Hopkins, Matthew (1640): *The Discovery of Witches: A Study of Master Matthew Hopkins, commonly call'd Witch Finder Generall*, ed. Montague Summers, London, 1939, p. 34b.

hence excellent observers – were considered especially interesting and included in the minutes of the meetings. Thus, not infrequently a beefy sailor's story found its way into the annals of educated society.[8] These are only some of the ideas that were used in Bacon's time to explain the relation between humans and their environment.

6.4. Hegel: The Dynamics of Concepts

[37] Modern science and modern philosophy of nature have stimulated the dynamics of concepts and perceptions without awareness of this process. Even Bacon desired firm concepts and firm sensations. Yet he noticed that the existing concepts and perceptions did not accurately represent the environment; hence he proposed replacing them with superior ones. Humans needed to change in order to be in harmony with the world. Bacon did not describe the dynamics of the concepts and perceptions in the transitional period. But we do encounter such a description in Hegel's work.

According to Hegel, who followed Aristotle, motion was a basic phenomenon in the world. However, for him motion was no longer a transition of forms that do not undergo any changes themselves. It involves the forms themselves, including the conceptual apparatus that we use to articulate the world, and in all of these cases motion is subject to the same basic laws. Stability in our concepts and ideas does not show that we have found the truth. Such stability is evidence that we have not succeeded in getting beyond a random state in our knowledge process to reach a higher state of consciousness and understanding. The question even arises whether we actually possess any knowledge in such a state. We adopt the existing categories, we learn the available alternatives until we can repeat them without any difficulty, and our thinking loses its spontaneity; it turns into "bestial staring at the world" (Hegel 1802b: 319/1802a: 202): "The more stable and splendid the edifice of the intellect is the more restless becomes the striving of the life that is caught up in it as a part to get out of it, and raise itself to freedom" (1801b: 90/1801a: 13). Any obstacle that we encounter, any seeming imperfection opens up new possibilities and temporarily provides the mind with freedom and spontaneity, which are, after all, its very own characteristics: "Process becomes converted back to praxis, the patient becomes an agent" (Laing 1967). Yet total freedom never arrives through the

---

[8] See the example in Weld (1848: I, 107f.).

185

common forms of thought. For all change, however drastic it may be at the beginning, leads to a new system of *stipulated* categories. "Thoughtful reason" still sharpens the existence of diverse terms, of a manifold, "to essential distinction, to opposition" (Hegel 1813b: 384/1813a: 61).

The methods of classical physics extend these "bad practices of reflection, which demands comprehensibility, but for that it presupposes its fixed categories and is thereby assured from the start to be forearmed against the answer to what it asks" (1812b: 72/1812a: 82) to the long presupposed and unanalyzed opposition between a subject and a very distinct world of oppositions, despite the fact that they allow for movement in special cases (1801b: 91f./1801a: 14). The following assumptions come into play here:

> the object is complete and finished all by itself and, for its actuality, can fully dispense with thought; thought, for its part, is something deficient and in need of a material in order to complete itself, and also, as a pliable indeterminate form, must adapt itself to its matter. Truth is the agreement of thought with the subject matter, and in order to produce this agreement – for it is not there on its own account – thought is expected to be subservient and responsive to the subject matter.
> (1812b: 24/1812a: 25).

> If thought and appearance do not completely correspond with each other, we have a choice, initially, of which of them to regard as the deficient one. [In scientific empiricism,] so far as it concerns the rational, the defect is shifted onto the thoughts; they are found to be unsatisfactory because they do not match up with what is perceived, or with a consciousness that restricts itself to the range of perception [so that] these thoughts are not to be found in a consciousness of this sort.
> (1830b: §47, p. 90f. /1830a: §47, p. 71)

Thus far about the standard conceptions of the relationship between subject and object (1812b: 24/1812a: 25). They are an integral component of the mechanism that Descartes introduced and Newton developed further, and they are responsible for the rigidity of classic science and its accompanying philosophy of nature that still remains despite the occurrence of quite drastic conceptual movements.

How can we overcome this rigidity? How do we obtain insights into the basic prerequisites not only of science, philosophy, and common sense, but also of our very existence as thinking beings? We have no insights as long as our views constitute an unreflecting, unvarying part of our lives. And yet if they change, isn't it then one person who begins the task of critique and another person who completes it? Problems such as this do not only arise if we pose the abstract question of the objective basis of all understanding in as general terms as possible, they also feature prominently in recent discoveries in the methodology, anthropology, and history of science. Hegel's cosmology contained a first answer to this question as well as a general theory of motion that leads us away from the cosmology of mechanism and toward a new and more flexible cosmology.

[38] Modern science and the philosophy that accompanies it and sometimes pushes it along trigger the dynamics of concepts as well as perceptions. This happens tacitly, since logic, methodology, and most scientists' and philosophers' talk about their professional activities continue to uphold the illusion of a stable basis and stable concepts. Moreover, not all concepts dissolve, and the ones that change soon come to rest due to the increasing success of the mechanistic world-view. Thus, modern thought about nature suffers from a threefold form of irrationality. We can remove this irrationality by admitting the dynamics of concepts as they actually occur, by trying to identify the laws governing these dynamics, and by setting in motion material that is still rigid or has become rigid again. This is why Hegel wrote that "the task is" to make "a fully ready and well-entrenched, one may even say ossified, material [. . .] fluid again, to revive the concept in such a dead matter" (1813b: 507/1813a: 211). The point is to "suspend the rigidified opposition between subjectivity and objectivity; to comprehend the achieved existence (*das Gewordensein*) of the intellectual and real world as a becoming. Its being as a product must be comprehended as a producing" (1801b: 91/1801a: 14).[9] The suspension is brought about by reason, which opposes "ordinary common sense with its typical fixation of opposites," thereby "nullifying" rigid science and common sense (1801b: 101/1801a: 25). This nullification is not the conscious act of a thinker who makes a *decision* to eliminate a distinction in his or her domain. To be sure, thinking subjects may try to overcome the limits of their state of knowledge by suspending the opposition in their *consciousness*. However, this local and subjective act will effect the rebuilding of science and philosophy, that is, of objective thought, only if it takes place in a suitable objective environment and under suitable objective conditions.

Hegel's general theory of motion, his cosmology (which he developed primarily in his *Logic*), offers more detailed information about

---

[9] See Lenin's comment on a similar passage in his notes on Hegel's *Science of Logic* (1929: part III).

such conditions. According to this cosmology, every object, every concrete being, stands in relation to all other objects: "A determinate, a finite being, is one that refers to another; it is a content that stands in the relation of necessity to another content, to the whole world. As regards the reciprocal determinations that hold the whole together, metaphysics could make the basically tautological claim that if one speck of dust were destroyed the whole universe would collapse" (1812b: 62/1812a: 71). This relationship is not external. Every process, every object, every state *contains* a part of the nature of every other object, state, process, and so on.[10] Conceptually this means that the complete description of an object contains a self-contradiction. It contains elements stating what the object *is*; these are the elements used in common sense or recent scientific descriptions. But the description also contains elements stating what the object *is not*. These are the elements that science, mechanistic philosophy, and even common sense assign to the object's exterior completely separate from the object. And yet they belong to the object itself and are responsible for its intrinsic self-contradiction (1812b: 44/1812a: 53). We cannot avoid the contradiction by using different *words*. Whatever we talk about, it has to be isolated from something that is different from it at least in our thoughts; otherwise it would be pure being, which is the same as pure nothingness (1812b: 58/1812a: 67). As soon as the separation has been made, however, the contradiction between the inner and the outer arises, and it does so within the object itself according to the aforementioned basic principle of Hegel's cosmology. Hegel had a great talent for making visible the contradictions that arise when we undertake a detailed analysis of a concept to try to fully understand the subject matter that it describes. "Hegel analyses concepts that usually appear to be dead and shows that there *is* movement in them" (Lenin 1929: 110).

This leads us to a second principle of Hegel's cosmology. The movement of concepts is not just a movement of the *intellect*, starting with the analysis of certain conditions, then moving away from them, and eventually positing their negation. It is at the same time also an objective movement caused by the fact that every finite thing (process, state, etc.) has a tendency to highlight the elements of other things that it implies and to become what it is not. The object, full of "the unrest of the something [to be] in its limit" (Hegel

---

[10] Bohm's essay in *Scientific Change* (Bohm 1960) offers a modern application of this idea in physics.

1812b: 101/1812a: 115), strives "not to be what it is" (1804/5b: 35, 1804/5a: 31).

> When we say of things that they are finite, we understand by this that they not only have a determinateness [. . .] but rather that non-being constitutes their nature, their being. Finite things are [. . .] but the truth of this being is (as in Latin) their finis, their end. The finite does not just alter, as the something in general does, but perishes, and its perishing is not just a mere possibility, as if it might be without perishing. Rather, the being as such of finite things is to have the germ of this transgression in their in-itselfness: the hour of their birth is the hour of their death. [. . .] The finite thus does indeed let itself be submitted to flux.
>
> (Hegel 1812b: 101f./1812a: 116f.)

By moving beyond its limit, the finite object ceases to be what it is and becomes what it is not; it is *negated*. A third principle of Hegel's cosmology is that "negation is equally positive, or that what is self-contradictory does not resolve itself into a nullity, into abstract nothingness, but essentially only into the negation of its particular content [. . .]. Because the result, the negation, is a determinate negation, it has a *content*" (1812b: 33/1812a: 35f.). Conceptually speaking,

> It is a new concept but one higher and richer than the preceding – richer because it negates or opposes the preceding and therefore contains it, and it contains even more than that, for it is the unity of itself and its opposite. – It is above all in this way that the system of concepts is to be erected – and it has to come to completion in an unstoppable and pure progression that admits of nothing extraneous.
>
> (1812b: 33/1812a: 36)

This is an excellent description of the transition from Newton's to Einstein's concept of space, provided that we continue to use the *unaltered* Newtonian concepts. "It is clear that no expositions can be accepted as scientifically valid that do not follow the progression of this method and are not in tune with its simple rhythm, for it is the course of the fact itself. [. . .] In keeping with this method I remind the reader," Hegel continues in the same passage,

> that the divisions and the headings of the books, the sections and chapters given in this work, as well as the explanations associated with them, are made for the purpose of a preliminary overview, and that strictly speaking they only are of historical value. They do not belong to the content and body of the science but are rather compilations of an external reflection which has already gone through the whole of the

exposition, therefore knows the sequence of its moments in advance
and anticipates them before they are brought on by the matter at issue
itself.

(1812b: 34/1812a: 36)

This is an apt criticism of contemporary history of science
and of so-called rational reconstructions in philosophy of science.
Considering that a movement beyond the limits is not arbitrary but
aims toward "the end" of the object, it follows that not all of the
aspects of other things inherent in the object will be realized in the
subsequent phase.

> Negation in dialectics does not mean simply saying no, or declaring
> that something does not exist, or destroying it in any way one likes.
> [. . .] Every kind of thing [. . .] has a peculiar way of being negated in
> such manner that it gives rise to a development, and it is just the same
> with every kind of conception or idea. [. . .] This has to be learnt, like
> everything else.
>
> (Engels 1894b: 85f./1894a: 173f.)

It also has to be learnt that the "negation of negation" does not take
us further away from the starting point but brings us back to it (Hegel
1812b: 110/1812a: 107). This is "[a]n extremely general – and for
this reason extremely far-reaching and important – law of development of nature, history, and thought; a law which [. . .] holds good in
the animal and plant kingdoms, in geology, in mathematics, in history
and in philosophy" (Engels 1894b: 85/1894a: 172f.).

> Let us take a grain of barley. Billions of such grains of barley are
> milled, boiled and brewed and then consumed. But if such a grain
> of barley meets with conditions which are normal for it, if it falls on
> suitable soil, then under the influence of heat and moisture it under-
> goes a specific change, it germinates; the grain as such ceases to exist,
> it is negated, and in its place appears the plant which has arisen from
> it, the negation of the grain. But what is the normal life-process of this
> plant? It grows, flowers, is fertilised and finally once more produces
> grains of barley, and as soon as these have ripened the stalk dies, is in
> its turn negated. As a result of this negation of the negation we have
> once again the original grain of barley, but not as a single unit, but
> ten-, twenty- or thirtyfold. Species of grain change extremely slowly,
> and so the barley of today is almost the same as it was a century ago.
> But if we take a plastic ornamental plant, for example a dahlia or an
> orchid, and treat the seed and the plant which grows from it accord-
> ing to the gardener's art, we get as a result of this negation of the
> negation not only more seeds, but also qualitatively improved seeds,

190

which produce more beautiful flowers, and each repetition of this process, each fresh negation of the negation, enhances this process of perfection.

(Engels 1894b: 81f./1894a: 166)

It is obvious that I do not say anything concerning the particular process of development of, for example, a grain of barley from germination to the death of the fruit-bearing plant, if I say it is a negation of the negation. [. . .] When I say that all these processes are a negation of the negation, I bring them all together under this one law of motion, and for this very reason I leave out of account the specific peculiarities of each individual process. Dialectics, however, is nothing more than the science of the general laws of motion and development of nature, human society and thought.

(Engels 1894b: 85/1894a: 173)

Thus far we have discussed concepts and objects as separate from each other. We have determined similarities and relations: every thing (process, state, and so forth) *contains* elements of other things, it develops by taking on such foreign elements, and it eventually attempts to *return* to itself. According to this view, the concept of a thing therefore implies contradictory elements. It is negated, and it develops in a manner precisely corresponding to the way in which the thing moves. This account has a serious flaw: it describes thinking as entirely subjective, while the world of objects is something independent that undergoes changes independently of the dynamics of concepts. This dualism has to be replaced by a view in which the subject itself is regarded as only a developmental stage of being, so that even the concept "the general development of nature" does not *participate* in the general development of nature. "Life," for example, "is the stage of nature where the concept comes on the scene, but as a blind concept that does not comprehend itself, that is, is not thought" (Hegel 1813b: 517/1813a: 224). As part of the natural behavior first of an organism and subsequently of a thinking being it does not just reflect a nature "all manifoldness [of which] falls outside it" (1813b: 519/1813a: 227), but is something "merely subjective and contingent" (1813b: 671/1813a: 408). It is not "only a concept" (1813b: 518/1813a: 225) but is part of the general nature of all things; that is, it tends toward being the end point of the movement of a certain thing, it contains an element of everything else, and its development eventually leads to "the unity of the concept and objectivity" (1813b: 671/1813a: 408). "That actual things are not congruent with the idea ['read: man's knowledge' – Lenin 1929: 194] constitutes the side of their finitude, of their untruth, and it is

191

according to this side that they are objects, each in accordance with its specific sphere, and, in the relations of objectivity, determined as mechanical, chemical, or by an external purpose" (Hegel 1813b: 672/1813a: 410). In this stage "nothing can be more harmful and unworthy of a philosopher than the vulgar appeal to experience, which supposedly contradicts the idea" (1813b: 671/1813a: 408). If something does not correspond to its concept then it needs to be guided toward it until the "identity of concept and thing" has been re-established (1813b: 521/1813a: 228).

This is how Hegel accommodated some of the problems caused by the rise of modern science and modern philosophy of nature in the context of an original new cosmology comprising matter, individual minds, and society. Hegel's philosophy brought to an end the period of mechanism that Descartes had started. This accomplishment has been less definite and less clear in the natural sciences. I shall give a brief overview.

6.5. Newton, Leibniz, Mach: Problems of Mechanism

[39] Henry More was strongly opposed to the Cartesian identification of space and matter. His arguments, which are in part of a physical nature, are very compelling. However, his main objective was to reintroduce permeable, indivisible, and yet extended entities – that is, spirits – into the world (there was no place for spirits in the Cartesian world). His description of such entities perfectly matches light[11] as well as the modern notion of a *field*. More's space, separate from matter, was an infinite, homogenous attribute of God. The space employed by *Isaac Newton* had very similar qualities. It was infinite, homogenous, a substance, and contained (without being identical to) all matter, which in turn consisted of small, hard atoms. Newton had powerful arguments against the fullness of the Cartesian world: the ease with which comets shoot in all directions through the solar system without bothering in the least about the Cartesian vortexes, as well as the impossibility of explaining the laws of planetary motions in terms of hydrodynamics. Newton used gravity instead of vortexes. Yet he was reluctant to conceive of the gravitational forces as new physical principles; as a matter of fact, he regarded the existence of physical principles with the properties of gravitational forces as downright absurd. He

---

[11] This was a continuance of medieval light astrology and light metaphysics!

suggested – and Bentley took full advantage of these suggestions – that gravity had to be caused by a spiritual being. Matter in the form of impenetrable atoms, forces most likely caused by spiritual beings, and both embedded in absolute space and absolute time, each of which is void, but real – this was Newton's universe. "A Frenchman," wrote Voltaire in his *Letters Concerning the English Nation*, "who arrives in London will find philosophy, like every thing else, very much chang'd there. He had left the world a *plenum*, and he now finds it a *vacuum*" (Voltaire 1733).[12]

Now, at the beginning of the eighteenth century this worldview was attacked from two sides: by *Leibniz*, who to some extent took up the Cartesian arguments but at the same time also developed his own theological, metaphysical, and philosophy-of-nature-related ideas, and by *Berkeley*, the astute precursor of modern positivism. Both thinkers were guided by the concern that Newton's philosophy of nature was threatening religion and would lead to materialism. Apart from this motive the two thinkers' approaches were very different. Berkeley's arguments were essentially epistemological. He did not try to show that Newton's *principles* did not exist; he went much further and challenged the meaningfulness of the very *words* used to describe these principles. For him only words designating sense impressions could be meaningful. And neither gravity nor matter is directly related to sense impressions. These words can be meaningful only as shorthand for a certain class of sense impressions. Ontologically speaking this means that only sense impressions exist. Gravity, matter, and atoms do not exist, though the words "gravity," "matter," "atoms" function as organizing tools to economically represent sense impressions. This method of "semantic analysis" is bound to eventually dissolve philosophy of nature, and that is indeed what it goes on to do in the nineteenth century (Mach). Berkeley's doctrine of space is interesting independently of his positivism (though the two positions are linked to one another). His little work "On Motion," in which he developed this theory of space, should be regarded as a masterpiece: if the coordination of sense impressions is the objective of science then it can make use of spatial relationships only if these are regarded as nothing but relations between perceptible bodies.

> For up, down, and all place or regions to the left or right are based on some relation, and they necessarily presuppose some body that is moved. Thus if one assumes that all bodies are reduced to nothing

---

[12] Notably, these Voltaire letters first introduced Newton's worldview, together with Locke's philosophy, in continental Europe.

and, for example, that a single globe exists, it would be impossible to conceive of any motion in the latter [. . .]. Thus, since motion is by its nature relative, it cannot be conceived unless other correlative bodies are given.

(Berkeley 1721: §§58, 59)

Even more radical was Leibniz's defense of the relativity of space and time. "No matter without space" meant for him that space and even time were nothing but "relational orders" among existing things: "But things being once resolved upon, together with their relations, there remains no longer any choice about the time and the place, which of themselves have nothing in them real, nothing that can distinguish them, nothing that is at all discernible" (Leibniz 1715–16: §57, p. 342). (Similar passages can be found in Descartes, for whom the interregnum between two worlds of which one is destroyed and the other created "somewhat later" is atemporal.) The arguments Leibniz offered to support his position were different from Berkeley's: if there were something like void, absolute, homogenous space then there wouldn't be a sufficient reason for God to create the first piece of matter here and not there. But the principle of sufficient reason is for Leibniz the foundation of all philosophy. According to this principle every judgment must have a reason from which it can be inferred.

Leibniz' and Berkeley's arguments are very plausible. And yet we should not overlook that neither of the two philosophers succeeded in explaining the inertial forces – Newton's most important argument for the existence of absolute space – on the basis of relations between physical elements. Only Einstein provided such an explanation, not by rejecting but by modifying Newton's idea of absolute space that is distinct from matter and yet no longer unaffected by it. This was also the beginning of the dissolution of mechanism with its rigid elements that could enter only into external interrelations.

The discussion of absolute space, in which Leibniz actually got involved against his will, was only a part of his criticism of Newton's worldview. Leibniz complained that Newton required divine intervention in order for his cosmos to keep operating. A physical argument used by Newton's supporters in favor of the assumption that the universe would come to rest if a supernatural force did not interfere either constantly or at least at certain intervals was the subsiding of impulse force in non-elastic impacts. Leibniz, just like Descartes, required that the world be able to be explained according to its own laws. Leibniz, and even more so Descartes but not Newton, were the precursors of Laplace's world machine. For Newton, space and time were attributes of God. And the famous *Scholium Generale*, which

194

he added on to the second edition of the *Principia* and where he discussed his metaphysical ideas, ends with the following thoughts:

> And now we might add something concerning a certain most subtle Spirit which pervades and lies hid in all gross bodies; by the force and action of which Spirit the particles of bodies mutually attract one another at near distances, and cohere, if contiguous; and electric bodies operate to greater distances, as well repelling as attracting the neighbouring corpuscles; and light is emitted, reflected, refracted, inflected, and heats bodies; and all sensation is excited, and the members of animal bodies move at the command of the will, namely, by the vibrations of this Spirit [. . .]. But these are things that cannot be explained in few words, nor are we furnished with that sufficiency of experiments which is required to an accurate determination and demonstration of the laws by which this electric and elastic Spirit operates.
>
> (Newton 1713: 507)

Newton's God intervenes in this world just as the biblical God intervened during the first six days of creation, and Newton considered this intervention necessary for the preservation of the cosmos. Leibniz' and Descartes' God, by contrast, is "the Biblical God on the Sabbath Day, the God who has finished his work and who finds it good, nay, the very best of all possible worlds" (Koyré 1957: 217). In the eighteenth century, people became convinced that philosophy of nature and physics no longer required the hypothesis of divine intervention (Laplace's reply to Napoleon – "I had no need of this hypothesis" – had precisely this meaning). This created a new stage in the mechanistic worldview.

[40] In this new stage, which soon became generally accepted, gravitational and other forces, which Descartes had still regarded as effects of the direct impact of particles, were conceived of as new and independent physical principles. Atoms and forces in a void, absolute space – this was the slogan of *mechanistic philosophy of nature in the eighteenth and nineteenth centuries*, the third major philosophy of nature after Aristotle and Descartes. In the nineteenth century, mechanism was the credo not only of physicists (with the exception of the genial Faraday and his successors) but also of biologists. The impressive development of all sciences in that century – for example, Helmholtz' explanations of numerous psychological phenomena in materialistic terms – provided mechanism with a vast number of arguments. But at the same time it laid the groundwork for mechanism's dissolution. Contemporary philosophy of nature is the result of this dissolution, which has been very complex and is not yet complete, and which combines radically diverse interests, particularly

195

those theories of the mind that had been so strongly opposed first by Cartesianism and then by mechanism. Even mythical forms of thought returned, namely in connection with recent developments in systems theory, to some extent stimulating scientific research and giving its results a new appearance. Hegel was the only philosopher who, anticipating and actively supporting the dissolution process, put together a comprehensive philosophical system for the next stage. We otherwise find only fragments, which may be quite interesting but are difficult to combine coherently into one new system. Let us take a closer look at some of them!

This aspiration meets an obstacle right from the start, namely the fact that in the nineteenth century, epistemological arguments start playing a significant part in philosophy of nature.[13] This had already started with *Kant*. Kant's major philosophical question was not "What constitutes the world?" but "What can we *know* about the world?" He was deeply convinced of the accuracy of Newton's worldview (which he considerably expanded with his cosmological hypothesis) and his objective was to justify it. And his justification was epistemological, not in terms of philosophy of nature. The principles belonging to general science, such as the principle of causation and the principles of conservation (and later on almost the entire Newtonian theory), were for him absolutely true because our very intellect thinks in the Newtonian way. Thus, they were first of all rules of our understanding, and *nature* was (chaotic) *sensation*, which had to be arranged into an order by means of these rules of the understanding. This explained why we are able to comprehend nature: its general traits are our own work. According to Ernst Mach, who proceeded in an even more radical manner, the mechanistic notion of nature had no foundation in experience and needed to be eliminated. It was not drawn from a classification of the elements in the world that would have been guided by the relations among these elements, hence it was based not on a *natural classification* but on random categories such as subject, object, understanding, world, and so forth, *that would first require examination*. Such examination could yield the result that elements are linked otherwise than is generally assumed in physics and physiology, thereby suggesting a new, homogenous (monistic) account of all events. Mach's ideas did away not only with the general notion of nature but also with the special

---

[13] The use of epistemological arguments to criticize positions in philosophy of nature, as well as the subsequent partial dissolution of philosophy of nature, are quite comparable with the sophists' criticism of Pre-Socratic philosophy of nature and the subsequent partial dissolution of that movement.

notions of substance, absolute space, absolute time, and atoms, and even with the core element of the mechanistic worldview: mechanics itself. In this Berkeley had been his predecessor, since he was the first to condemn Newton's philosophy of nature from a positivist point of view and, as we saw, also to deliver a compelling criticism of absolute space, absolute time, gravity, and matter. Mach differed from Berkeley in that he wanted to see the nature of elements determined by scientific research itself. According to him, his conjecture that the elements were sense data was just a *hypothesis* subject to possible future revision. Mach's criticism, like Kant's philosophy before him, stimulated the *epistemological* dissolution of mechanism. (Hegel's criticism was of a cosmological nature.)

6.6. Einstein, Bohr, Bohm: Signs of a New Era

[41] Mechanism also faced physics-based challenges in the nineteenth century. Maxwell's electrodynamics, found accurate by Hertz' research, employed the notion of a *field*, which did not have a rational *and unambiguous* mechanistic interpretation (Poincaré's theorem). Thermal phenomena display a directionality that is foreign to mechanics (temperature differences offset one another but they do not come about without external factors) and, as can be shown, actually contradict mechanics (Poincaré–Zermelo paradox; Loschmidt's reversibility paradox). Boltzmann and Einstein managed to solve this contradiction, but only at the price of introducing new, nonmechanistic principles such as objective probabilities. Mechanism experienced a direct hit, however, from *special relativity theory*, according to which qualities ascribed in mechanics to matter, space, or time (such as inertia, spatial or temporal extension) should be considered *relations* between events and coordinate systems, so that they can change corresponding to changes in the dynamic state of the respective coordinate system, and every form of energy, including the energy of gravitational fields, possesses inertia and thus a material nature. Yet relativistic *space-time* is still absolute, and the same goes for the relations between events in space-time. Furthermore, it possesses a well-defined structure: events that can be linked by light signals are simultaneous, events in the bottom part of the light cone belong to the past, and events in the top part of the light cone belong to the future. (In Newton's space-time the light cone collapses into a double-layered space plane. Thus, the structure of Newton's space-time consists of a vertical fraying and a horizontal layering.)

197

Overall, special relativity theory acknowledges three types of objective entities: space-time, matter, and (electromagnetic and gravitational) fields. Roughly speaking, the transition to general relativity theory consists of a fusion of space-time and gravitational field. To accomplish this fusion the general theory does not use the level space-time of special relativity but the curved Riemannian one. We all know that a spatiotemporally curved (that is, accelerated) motion reveals the inertial properties of matter: forces appear (centrifugal forces; Coriolis effect). In this way Newton already tried to distinguish between relative and absolute motion. Thus, a curvature of space-time that cannot be eliminated everywhere by means of coordinate transformation generates "genuine" forces (as opposed to "mock" forces, for example, centrifugal ones) such as gravity, which is thereby revealed as a phenomenon of inertia. And the curvature in turn depends on the distribution of matter. It is notable that this curved, non-homogenous, and non-isotropic space-time of general relativity theory does have an independent or, if you wish, "absolute" role; Newton's absolute space and absolute time are not eliminated by relativity theory, they are just fused and redesigned as Riemann's space-time, which takes on the role of the old ether. "To deny the ether is ultimately to assume that empty space has no physical qualities whatever. The fundamental facts of mechanics do not harmonize with this view" (Einstein 1920b: 16/1920a: 12).

An attempt was made to include the electromagnetic field in the spatial structure, but it was not very successful (Weyl; Kaluza). A second attempt turned out to be more significant, since de Broglie and his students repeated it a few years ago in a different context. Einstein and Grommer attempted to interpret the particles of matter as singularities in curved space-time, thereby fusing space and matter into one unit. This attempt, which advanced to the point where the equations of motion could be derived from the field equations (Einstein, Infeld, and Hoffmann), leaves only two entities as basic principles, namely space-time and the electromagnetic field. This dramatic simplification can be regarded as the high point of classical *physics* – the kind of physics that conceives of the world as a being independent of the subject, and that aims to explore the fundamental principles of this world. For Newton these principles were (divine) space, time, (atomistic) matter, and gravity. These were later supplemented by fields of two different kinds: the deterministic field (Faraday), to which various thinkers (Thomson; Lorentz; Lamour) attempted to reduce matter, and the probability field (statistical mechanics). Einstein combined space and time into space-time, to which he reduced gravity and perhaps even the electrical field as well as matter.

The curvature of space-time also enabled us to address the *cosmological problem* in a new way. The infinite Newtonian world was unstable (something that was apparently first shown by Heckmann), and yet a finite collection of particles of matter in Newton's space will dissipate (Einstein's evaporation objection). Curved space-time is also unstable, but it allows for meaningful dynamics, is able to pulsate, and allows us to form the image of a world that is unlimited and yet finite, and that also contains only a finite collection of particles. These possibilities have stimulated our imagination and led to various models of the world, all of which reflect the contents of our experience more or less adequately but otherwise have very little in common. This is an entirely new stage for astrophysics and cosmology.[14] But all of these attempts, together with their underlying proud edifice of Einstein's philosophy of nature, have been seriously challenged by our new understanding of matter associated with the name of *quantum theory*, as well as by a philosophical movement supported by that understanding's inventors, an understanding that demands nothing less than the abandonment of Einstein's most fundamental principles, the principle of objectivity and the principle of determinism. Let us take a look at this new development.

[42] The ideas of philosophy of nature that accompanied and promoted the development of modern microphysics are available to us in an "original" version and a "popular" version, the latter of which mingles elements of the former and establishes a rather superficial connection to contemporary positivism. The original version is basically *Niels Bohr*'s philosophy. Bohr's philosophy also has two versions: one that is more abstract and speculative, and another that is more physics-based and concrete.[15] The abstract version is closely related to Mach's philosophy, though it also incorporates characteristics of Kierkegaard's thought, with which Bohr was made familiar through Harald Höffding. It acknowledges that the delineation of an object is always to a certain extent arbitrary and that no presentation of our knowledge can be anything but a *transitory stage* based on arbitrary delineations. There are no "results," though every discussion reaches a certain stage at a certain time, simply because at that time the future has not yet become the present. This abstract idea

---

[14] A complete overview is offered in Misner, Thorne, and Wheeler (1973).
[15] For the following see Feyerabend (1968, 1969a).

explains a feature of Bohr's works that at first glance seems rather mysterious: *all* of them start with an outline, as complete as possible, of the development of a problem, and *all* of them end more or less inconclusively or arbitrarily with a catalog of the most recent results of the discussion.

> He would never try to outline any finished picture, but would patiently go through all the phases of a problem, starting from some apparent paradox, and gradually leading to its elucidation. In fact he would never regard achieved results in any other light than as starting points for further exploration. In speculating about the prospects of some line of investigation he would dismiss the usual considerations of simplicity, elegance or even consistency with the remark that such qualities can only be properly judged after the event.
>
> (Rosenfeld 1967: 117)[16]

The affinity with Kierkegaard is obvious:

> While objective thought translates everything into results and helps all mankind to cheat, by copying these off and citing them by rote, subjective thought puts everything in process and omits the results; partly because this belongs to him who has the way and partly because as an existing individual he is constantly in process of coming to be, which holds true of every human being who has not permitted himself to be deceived into becoming objective, inhumanely identifying himself with speculative philosophy in the abstract.
>
> (Kierkegaard 1846: 68)

So much for the abstract version of Bohr's philosophy. Originally this abstract version was only loosely linked to Bohr's research in physics. But as his research progressed Bohr discovered to his surprise that the hard facts of physics *forced* the observer to draw arbitrary demarcation lines and distinctions, thereby supporting the general philosophical ideas. This discovery led Bohr to formulate his idea of *complementarity*, which assigns the subject a crucial role in the shaping of natural phenomena. It has this role not because of the abstract-subjective quality of our knowledge but because of certain regularities in the physical body that generates this knowledge.

It soon turned out that the elements of matter were not as simple as had been assumed. The atoms, the atomic cores, and even the elementary particles all have a *structure*. For one thing, this multiplies the *number* of elementary particles that make up matter (today we

---

[16] The collection containing this contribution includes a wealth of material on Bohr's philosophy.

are acquainted with about a hundred of them) as well as of fields (today we recognize four types of fields: gravity, electromagnetic field, strong interaction, and weak interaction). But it was especially our knowledge of the *nature* of these elementary particles that pushed for radical changes to the classical position. Since about the middle of the nineteenth century it had seemed to be firmly established that *light* is a wave movement emanating from the light source in concentric spheres. Experiments conducted in the first third of the twentieth century then resulted in the conjecture that the *energy* transported in the light field has to be concentrated in packages, and that the interaction of light and matter has to be accompanied by a directed exchange of impulse (the Compton effect). Thus light appears to be some kind of particle scattering. But this conjecture cannot explain the phenomena associated with interference. And this is how our very specific experiences of light (and, as it turns out, of elementary particles in general) fall into two groups. The experiences of the first group can be precisely explained in terms of wave theory, but they contradict the assumption that they involve particles. The experiences of the second group can be precisely explained in terms of the particle model, but they contradict wave theory. This constitutes the dualism of light and matter. The crucial issue here is that the two sides of radiation and matter correspond to two sides – or better, two halves – of Einstein's space-time. The dualism of particle model and wave model implies a dualism of configuration and motion such that the former is determined by positions in space-time and motion by the momentary direction of the paths. The two cannot exist together, and space-time dissolves into directional space and configurative space. The notion of path becomes inapplicable (other than as an often rather rough approximation). And this means, as Poincaré had noted very early on, the dissolution of space-time and of motion as conceived in classical physics.

Now we might expect this dissolution to be followed by a synthesis based on a new and more comprehensive principle. Such a principle would conceive of particles and waves as two different sides of a more abstract entity, giving rise to the conjecture that this dualism forces us to accept yet another general account, and an objective one at that, of space, time, matter, and motion. The majority of contemporary physicists reject such an idea. "It would be a misconception," wrote Niels Bohr, "to believe that the difficulties of the atomic theory may be evaded by eventually replacing the concepts of classical physics by new conceptual forms" (Bohr 1934 16). We have to be content with the *ruins* of the classic worldview, which from now on function as

201

elements of a new worldview. These ruins have much in common with Mach's elements, for they are the result of a synthesis between "objective being" and "subjective thought"; but as a result of a long period of research they are also more determinate. It is *knowledge* and not, as in Berkeley, a dogmatic starting point that determines the structure of the elements and their relation to the subject.

In this new world, determinism is no longer fully valid (determinism was linked to the classic concept of a spatiotemporal path, which was abandoned and not replaced). Furthermore, the world is no longer independent of the observer: whether the wave or the particle model is applicable depends on which experiments are conducted, and since there is no underlying "deeper reality" to either of the two models, it is no longer possible to objectify the relationships involved in this dependency, as had been done in relativity theory. This also distinguishes Bohr's ideas from those of Hegel, who, while dealing with the separation of subject and object dialectically, nonetheless accommodated it in a universal conceptual framework.

For some time Einstein and Schrödinger were the only physicists who stood against this new conception of nature. Beginning in the 1950s a countermovement took shape that aims to return to the clear and objective ideas of classical physics and rejects the solution of difficulties in physics by way of "recourse to epistemology" (Schrödinger). This countermovement includes, on the one hand, ideas developed in quantum electrodynamics, according to which we should conceive of the world as filled with fine ether capable of existing in various excited states. These excited states are identified with elementary particles. There have been attempts to develop a discrete theory of space and time. On the other hand, we also have a genuine "counterrevolution" associated with the names of Bohm, Viglier, and de Broglie, and which had long been in the making in Russia as well. These physicists deliberately placed the idea of an objective description of nature at the top of their preferences, without intending to repeat the static ideal of objectivity that Hegel criticized in such a clear fashion (both Bohm and Viglier are dialectic materialists, and Bohm, in addition, is very familiar with Hegel).

According to Bohm all laws that we discover in a certain domain, and all distinctions that we make, are nothing but approximations whose validity has to be determined through subtler investigations. The fluctuations revealed in such investigations that make the aforementioned laws to some extent "indeterminate" can be explained at a deeper level, where new interactions between elements as well as new effects on the observing system – including the observer's thinking – are taken into consideration. The world is divided up into an infinite number of such layers that possess relative independence, though they also mutually affect and interrupt one another. Each indeterminacy can be explained via recourse to a deeper layer at which new quantitative and qualitative indeterminacies subsequently occur. Some regularities are applicable to all levels, and they enable Bohm to deduct the quantum conditions (which have universal validity as well) by means of simple observations of time lapses in the elements of a particular layer (each element possessing an inner time comparable to the beat of a pulse, which in turn is the effect of a superposition of the pulse beats of the infinite number of elements at the deeper levels). The system of levels itself can be explained in terms of set theory, which can also serve to determine local dimension values. Thus, the concept of *motion* (which no longer is continuous, though it still satisfies the ergodic theorem) becomes the basic concept, from which the concept of *extension* is derivative. Positive results that would lead beyond the familiar are not yet available and cannot be expected soon due to the difficulty of the entire project.

[43] The gradual replacement of mechanistic approaches in physics has led to revolutionary developments in other disciplines as well. The triumph of Cartesianism pushed aside not only certain theories but also a large number of obvious facts. This includes all those facts supporting an independent existence of the soul, which is not easy to explain in mechanistic terms, or the existence of mental powers that are independent of matter. There were numerous facts of this kind in the fifteenth century. These were collected in Sprenger's and Kramer's *Malleus Maleficarum* (1486), which explained them in terms of devils and demons. The triumph of mechanism, especially in its Cartesian form, continued to obstruct the development of a rational psychology for centuries, and it was not until the twentieth century that holistic approaches and related research programs came into their own again. Without a doubt, this was in part facilitated by developments in modern physics as well as by modulator theory, which has provided us with concepts for a holistic description of complex systems. The recognition of the importance of interaction and the role of subjective factors also went on to affect sociology, so that a crucial role in the formulation of its knowledge was *granted* and, as in the case of Brecht's theater and in psychoanalysis (up to Ronald D. Laing's approach), even *imposed on* the subject. This has also opened the door for a return of mythological forms of thought, which seemed to have

abandoned human thought forever with Parmenides. And with that we stand at the beginning of a new period whose traits, due to the manic professionalism in almost all disciplines, are still difficult to discern at this time.

— 7 —

CONCLUSION[1]

This completes my account of the first phase of Western philosophy of nature, namely the phase during which thought is separated from intuition in order to impose on the latter, as well as on life in general, its laws – or whatever are regarded as its laws – from the outside. Western philosophy's peculiarity lies precisely in this: human beings are separated from nature as well as from their own immediacy; they regard themselves as something alien which they attempt to grasp with the help of something else that is alien and has just been discovered, namely thought. Plato's call for a rebuilding of astronomy based on pure principles of *thought*, while "let[ting] be the [phenomena] in the heavens" (*Republic* 530c), is an excellent example of this approach, without which science as we know it would not have developed in the first place. It is this separation of thought and intuition, this conception of a *theoretical* science, which distinguishes Greek astronomy from the highly interesting and advanced Babylonian astronomy. No one would have been able to foresee the consequences of this separation; no one could have predicted that this approach would lead to any significant results, or to a self-contained, flourishing form of life, and indeed there are areas in our culture where the pursuit of a theoretical science has so far achieved nothing but empty talk and useless fights. Thus, it is our task to understand the contingencies that helped the endeavor to succeed, and to precisely observe the limits where success turns into babble.

My account of the first phase of Western philosophy of nature

---

[1] [Editors' note: Though the following summary covers only the first five chapters and is consistent with the original plan of distributing the topics of chapter 6 over two subsequent volumes, Feyerabend titled this text as chapter 7, numbered it with consecutive page numbers, and placed it after the sixth chapter at the end of the typescript.]

is by no means complete. I left out the Pythagoreans, who tried to bridge the chasm between thought and immediacy with new forms of life (Burkert 1962), and who were a significant – though not yet fully understood – factor in the development of Greek philosophy of nature. I also left out Heraclitus, who bequeathed to Plato some of the latter's most interesting problems. I failed to provide a detailed account of the cosmologies of Parmenides, the atomists, Empedocles, Anaxagoras, and many other influential thinkers. I also failed to provide a detailed analysis of the rational tools used in tragedy, and in other forms of life today classified as arts, in order to clarify familiar but no longer tolerated circumstances and to expose them to critical scrutiny. And I failed to provide a systematic account of the transition to philosophy of nature. I gave my reasons for this last omission in the relevant chapter. Research of unknown facts and even the classification of familiar facts had been prevented in the past both by naïve naturalism and by rash "explanations." Thus, at this stage the only course is to suggest possible future developments that future research will have to combine in a unified process. In doing so the limits of the development process should be set as broad as possible. The development of philosophy of nature (as well as of the accompanying new forms of poetry, new forms of painting, and so forth) is intrinsically linked with a dramatic increase in self-consciousness or perhaps even with the discovery of a spontaneous self. Hence we need to take into consideration recent research in psychology on the development of self-consciousness in the attempt to *arrange* familiar material and as heuristic principles in the *discovery* of new material. Only a close collaboration between anthropologists, sociologists, psychologists, philosophers, astronomers, mathematicians, and classical philologists can yield significant results in this difficult area of research.

Here is my plan for the two volumes to follow. The *second* volume is dedicated to Plato, Aristotle, and the medieval period up to the Renaissance. I shall compare Plato and Aristotle – despite varying interpretations not totally misunderstood – with Christianity and with each other. Upon the emergence of philosophy of nature *Aristotle* remained the only thinker who attempted to reconcile the demands of thought with intuition in such a way as to erect a complete dwelling in which we humans can feel at home and in a familiar environment again. It is no exaggeration to say that Aristotle *constructed a new, rational myth* in which even the quirkiest excesses of reason do not oppose natural life but are fully in its service. *Plato* is very different. He, too, accommodates myth. Yet Plato's myth is not a *theory* permeated by reason and intuition, not a visible form of reason or some

206

kind of popular science. Rather, it is an *anticipation of reason's future achievements*. It is a bridge to an insight that cannot be obtained by reason, and thus a dynamic principle that drives humans on toward grasping a form of being that is never present and nevertheless not unattainably far off, either. *Platonic science* is an *open* science, as *Timaeus* and the changing accounts in the other dialogues clearly show, especially if we compare those accounts with the explanation in the *Seventh Letter*.

No one can explain today why the grand Aristotelian myth gradually started crumbling in the fifteenth and sixteenth centuries, and in what way the new sciences of the sixteenth and seventeenth centuries managed to draw people's attention. Despite the great differences in detail, however, the situation displays a surprising similarity to the emergence of Ionic philosophy of nature from its preceding worldview. Thus here, too, we will have to take a broader perspective; developments in art and politics will have to be considered and explained alongside purely scientific arguments.

This procedure will be presented in the *third volume*, which covers the period that leads to the present time (around 1970). Concerning the present itself, it contains the core – in part clearly noticeable and in part hidden – of a new philosophy of nature and a new science. Alongside the large mass of the orthodox scientific enterprise, which is gradually turning into a business pushed forward by unhappy, fearful, and yet conceited slave souls, a new enterprise emerges using instruments of scientific research not in order to build clear, objective systems but to create a *process* that fuses humans and nature into a higher (but not at all totalitarian) unit. This process does not deprive humans either of our freedom or of the degree of knowledge that we need to tackle our problems in an ever-changing social and natural environment. Nor do we obtain this knowledge by excluding other areas of humanity or raping our surrounding nature. Sympathy with this nature, an intuitive understanding of the diverse life that it contains, and full development of one's own personality are essential components of the new philosophical and mythological science, the still indistinct outlines of which can be seen on the horizon. It is one of the aims of this work to clarify the historical preconditions – discoveries and errors – of this science, thus accelerating its birth.

Paul Feyerabend:
Previously Unpublished Documents

Letter to Jack J. C. Smart, December 1963

Dear Jack,
thank you very much for your letter and your most stimulating paper. I very much enjoyed both and I learned quite a lot from the latter. As a matter of fact I am now planning (as a result of reading your paper) to write a paper on *Science and Common Sense*[1] which will proceed from the way in which you put the problem. I would have liked to write you a better letter, but being ill again I can neither use the pen nor think very well. Here, then, are a few scattered remarks for whose incoherence I must apologize.

I agree that the issue between Nagel, Sellars and "myself" ("myself" sounds funny to me – I have always thought that I was only explaining what any reasonable person would have in his very bones and what many scientists have practiced. Just consider how Galileo in his *Dialogues* transforms commonsense with the simplest and the most shrewd of arguments) may depend on empirical considerations. If it was indeed true, as Kant thought it was, that certain reaction patterns, or conceptual schemes are inborn then it would quite obviously be impossible to replace them by different points of view. Still I do not believe that this possibility which is not yet established as a positive

[*Translator's note: All of the documents in this section were written by Feyerabend originally in English.*]
[1] [*Editors' note: Though Feyerabend never published a work with this title, some of his thoughts on the relation between common sense and science can be found in his discussion of Smart in Feyerabend (1965).*]

211

fact, can be used as an argument against total replacement, or at least against the initial *attempt* at total replacement. Two reasons for this:

(I.) There seems to exist strong *prima facie* evidence *against* the existence of insuperable conceptual schemes. This evidence lies in the fact that different cultures have developed radically different "manifest images" to use a term I think is very fitting. This sociological fact should really be given more weight in philosophical arguments. To take only one example: if the Egyptian *notion* of mental events (not that they were *called* that) should turn out to be different from the contemporary Oxford notion, then the demand that materialism be capable of explaining the latter, but not the former, completely loses its point. And so do all other demands which require synonymy, or co-extensionality with the "ordinary" notions. Now I think I can explain why this variety of manifest images is given so very little weight. It seems to be generally assumed that all these different points of view, the point of view of the Zulu, of the Hopi, of the Ancient Greek etc. are basically identical (with Strawsonism?) and that the differences belong to some theoretical superstructure, as it were. Austin, with whom I had many arguments over this matter always seemed to take it for granted that people took tables much more seriously than they took daemons and that the usages connected with table words were a much more solid part of "the" common idiom than were the usages connected with daemon words and he thought that daemons, therefore, occupied a rather peripheral place in the manifest image. Sellars, too, did not seem to be willing to regard the Hopi point of view as a *genuine* alternative of our own common philosophy (in a discussion I had with him in 1957). For Kant and many other thinkers, myths were a matter of fantasy and had nothing whatever to do with forms of thought. Cassirer seems to have been one of the first thinkers who adopted a different attitude here, although he, too, regarded for example mythical space "as having a position in between perceptual space and space of pure knowledge" (*Philosophie der Symbolischen Formen*, Vol. II, p. 107).* Yet I think we shall have to admit, following a careful analysis of the structure of the worldviews which have been connected with different cultures, that they are genuine alternatives, genuinely different manifest images. I wish I could present some examples at least to make this assertion more plausible. I shall certainly do this in one of my future papers. I always wanted to write a paper about the nature of myth in order to show that they are fully

---

* [*Translator's note: Cassirer, Ernst (1925): Philosophie der Symbolischen Formen. Vol. 2, Berlin.*]

blown worldviews. All I can do here is quote some Nietzsche. In his earlier life Nietzsche was very much interested in the philosophy of nature and he intended at some time to write a systematic treatise on this topic. He adopted for some time Mach's point of view and he interpreted this point of view as showing that the world in which we live (our perceptual world) is a mental construction, and that different kinds of such mental constructions are possible. This brings him at some places rather close to Wittgenstein. Thus in his *Wahrheit und Lüge im Außermoralischen Sinn* we find the following passage: "The waking day of a mystically excited people as are the older Greeks is indeed more similar to a dream than is the day of a sober mind; and this is due to the always present miracle as assumed by the myth. When every tree can now talk as a nymph, or when a god can rob virgins under the cover of a bull, when the goddess Athena is suddenly seen in person, driving side by side with Pisistratos in a splendid carriage through the market places of Athens – and this the honest Athenian did in fact believe – then like in a dream everything is possible at any moment and the whole of nature surrounds man as if it were only a masquerade of the gods."[2] (Lynn White in an excellent paper in the Volume *Scientific Change* ed. Crombie has pointed out that the transition from this world view to the cult of Saints, who kept to themselves and left nature alone swept nature clear of all spirits and thereby made it accessible, "conceptually" as well as "manifestly" both to technological interpretation, and to a purely mechanical interpretation. It was a change in the *manifest image* that made mechanicism possible and thus finally led to a more modern kind of "common sense.")[3]

To sum up: the variety of points of view in different cultures must be taken seriously, it must be taken as an indication, that non-Strawsonian manifest images are possible, and that Strawsonism therefore cannot be inborn (the evidence referred to by you and by Chomsky is much too sketchy, too weak, and too prejudiced to show that it is inborn. It is interpreted in a certain manner, and it has not been obtained under satisfactory conditions. We are not interested in how people react in a social environment that has many Strawsonian features and is therefore liable to restrict the manifold of reactions they are capable of).

[2] [Editors' note: Nietzsche (1873a, 1873b).]
[3] [Editors' note: White, Lynn (1963): "What Accelerated Technological Progress in the Western Middle Ages?," in: A. C. Crombie (ed.), Scientific Change: Historical Studies in the Intellectual, Social and Technical Conditions for Scientific Discovery and Technical Inventions, from Antiquity to Present, London, pp. 272–91.]

There is still another point that is worth considering and this is the possible existence of archetypes. I am very much in sympathy with the idea of Hughlings Jackson that man contains a hierarchy of reaction patterns corresponding to a hierarchy of neurological levels. Primitive patterns are fairly universal, higher patterns may show great variation from individual to individual. Primitive patterns are usually overlaid, and to that extent suppressed by higher patterns, but they may occasionally emerge when the higher levels are either damaged, or temporarily out of action. Now it may well be that the primitive patterns in all human beings constitute a certain rudimentary worldview that is common to all. I am even convinced that this is the case. However such a situation, far from making this worldview the indispensable *foundation* of everything else would only occasionally make it raise its ugly head and would make it thereby *difficult*, though by no means *impossible*, wholly to live inside a more reasonable worldview. The existence of lower levels beneath higher levels shows that it is difficult to be civilized (in one's behavior, as well as in one's thought). The existence of higher levels, on the other hand, as well as the possibility to modify these higher levels shows that it is *possible* to be civilized. Moreover it is *desirable* to be civilized, hence the reference to the lower levels loses all its force. However quite apart from the *irrelevance*, in view of the existence of a hierarchy of higher levels, of any reference to the features of some lower level, it is very doubtful to assume that these lower levels might be Strawsonian. Lower levels are those which man has in common with the higher animals. Now although animals don't speak, there are still means of exploring their manifest image (which, I guess, would in any case possess features that are independent of speech). And it seems that this manifest image is not at all Strawsonian. (It gets worse if we go farther back on the evolutionary ladder – deep sea fish are most likely Cartesians interpreting bodies as moving modifications of space-matter). To conclude this first part of the argument: it seems very unlikely that God built man in Strawson's image. If he built man in any image at all, then the limits of *this* image have not yet been reached.

But (II.) let us now assume, contrary to what emerged as plausible, that Kant is right and that we *are* incapable of leaving the boundaries of a certain worldview. How do we find out that this incapability exists? And how do we discover what is it we cannot shake off? Kant and many contemporary philosophers seem to take it for granted that such boundaries can be explored *from the inside*. They assume that the insuperability of commonsense can be discovered by procedures which move wholly inside commonsense. They therefore advise us to wait until the boundaries have been found and not to try rushing off in all directions in a vain effort to leave behind what cannot be left behind. Now this I take to be a completely wrong approach. The attempt to find boundaries by exploration from the inside is the surest way to erect artificial boundaries, boundaries which do not correspond to the possibilities of human nature *in general* but only to the possibilities of a human nature that has been paralyzed by a philosophical doctrine. The only way to discover whether there are indeed restrictions to our way of seeing the world is to resolutely try many different approaches and also to try applying them on all levels, on the level of commonsense as well as on the level of theoretical reason. Only if such a procedure fails again and again, only then may we try to explain these failures by the hypothesis of an unalterable inborn worldview (but even here we must be careful. What has been impossible for 3000 years may become possible tomorrow. Even within the rather restricted worldview of science it took about 2000 years until the atomic theory could be finally regarded as satisfactory). However what has just been said means that even if Kant should in fact be right we could find this out only by first adopting my point of view. This *methodological* consideration shows that there exists an interesting asymmetry regarding the views of Nagel, Sellars, and myself. The asymmetry consists in the methodological necessity to *start* with my view. Nagel and Sellars can never find out whether I am not unduly expanding it. Hence, if I am right, and if there exists no limit to human reason, then obviously my view should be adopted. However if I am wrong, that is if there *do* exist inborn limits to human reason, then my view and my procedure must be adopted too, for this is the only way to discover the limits. Now there either exist limits, or there don't. Hence, my view must in any case be adopted. And considering the success of those who adopted it in the past I have all confidence that my view is not only the necessary first step, but is also right, or at least I have all confidence that it is not Strawsonianism which defines the limits of human reason.

My second point is connected with your remark that when describing macroscopic objects such as galvanometers the physicist MUST use Strawsonian terms (p. 12). I think that this remark is *unsupported, unsupportable* as well as *false*. Let me consider some possible arguments in its favor:

(1) The physicist must proceed in this way because that is the way he is built – *false* for reasons I have just explained. (This first argument has been used both by Heisenberg and by Weizsaecker for retaining classical language at the macroscopic level

and classical language is for them *refined* commonsense not *reformed* commonsense).

(2) He must proceed in this way because he does not know how to proceed differently – both *irrelevant* and *false*. Irrelevant, because a deficiency is never an argument in favor of remaining deficient. False, because in many cases more adequate terminology or, if the terminology is still missing, more adequate interpretations of the common words are readily available.

(3) Because ordinary language is perfectly adequate for describing macroscopic objects – both *irrelevant* and *false*. Irrelevant, because the adequacy of the ordinary idiom should not prevent us from replacing it by a different idiom if we can in this way create greater coherence of expression. False, because many descriptive statements of commonsense are false, if taken in their commonsense interpretation. Example: material objects, such as this table in front of me are supposed by commonsense to be observer independent in the sense that they change their shape only when interfered with physically. Sirius does not change its shape when I am sitting here and shaking my head. Relativity teaches us that the shape of *all* objects is observer dependent and it therefore teaches us that "this is a table" using "table" in the usual sense of this word is *always* false. Now there is no need to introduce new words, such as "time slices" etc. We may retain all the commonsense *words* – but we sure must change their *meanings*. Thus we shall continue to say "this milk is too damn hot," but we shall now think of *motion* and *not* of substance when thinking of heat.

(4) Because the ordinary man understands perfectly well what he means by his terms when saying "the milk is hot" and he does *not* mean "the average velocity etc. etc.." *Irrelevant*. As irrelevant as possible! What has the ordinary man's point of view got to do with the way in which the scientist is supposed to describe his galvanometer? Assume the ordinary man knows what he is talking about. Does this mean that what he is talking about does also exist? How would this argument look when applied to what a Zulu says about his milk? He presumably also knows what he is talking about (spirits, I guess). Does this mean these spirits exist? Every inquisitor knew "what he was talking about" when diagnosing possession, etc. etc. Nagel seems to use this argument. I cannot understand how he can believe for a second that it is relevant.

(5) Because without being related to the common expressions physics would not have any meaning. *False*. This argument assumes that theoretical terms obtain their meaning by being related to the

216

terms of some pre-existing commonsense. However how did these terms obtain their meaning? And if it is possible to teach small children such rather abstract terms as "table," "chair," "god," "duty" etc. etc., then why should it not be possible to teach them "electron" etc. etc.?

(6)   Because different descriptions (more especially – descriptions inconsistent with the commonsense descriptions) would be *false* (this is the paradigm case argument). *False.* Reason same as in (3).

Having gone over all these reasons (and I think there is nothing I have left out) I must say that they are a sorry sight indeed. Is it not time to give up this illicit love affair with commonsense and to proceed to a more reasonable point of view? All these reasons also proceed from the wrong end. The question is what kind of language will be adequate as an observation language for scientific theories. This question can be answered very simply. There are three requirements: They are:

(a)   the language must be observational. This is pragmatic, or a psychological condition requiring that the production of certain signs of this language (in the relevant circumstances) be a reliable *indicator* of certain features of the surroundings.

(b)   It must be *factually adequate.*

(c)   It must be *relevant* for the test of the theories under review, that is its statements must be derivable (with the help of initial conditions) from the theory.

Now I challenge anyone to prove that (a) & (b) & (c) taken together entail that we must be Strawsonians. Condition (a) may lead to certain existing languages where conditioning has indeed established the needed behavioral mechanism. We may discover that Ordinary English satisfies (a) and might therefore want to retain it (after all, conditioning a person to the proper response in front of observable objects is not a simple matter and there is no reason why one should make life more difficult than it already is). But note that even if Ordinary English were to satisfy (a) this would only be an argument for retaining a very small part of its "grammar," *viz.* the part that is connected with direct observation. The remaining "theoretical" part of the grammar may be completely inadequate for *describing* what it so well *indicates.* Thus the shouts of a savage may *signal* to us the rising of the sun. However it would be incorrect to assume that what he means to convey by his shouts *viz.* that the sun god has just been reborn is a correct statement. Hence only very little of Ordinary English is brought in by (a). Nothing has as yet been said about its meaning.

217

I think it is easy to see how the fact that part of Ordinary English conforms to (a) could create the impression that Ordinary English is alright, *or perhaps even required*, as an observation language; how it could create the impression that satisfaction of (a) guarantees satisfaction of (b) and (c) also. These are the steps of unreason: *first step*: You can still call a table a "table" (correct); *second step*: the meaning of an observational term is constituted by its application in concrete situations (incorrect – this is the kind of thing Wittgenstein intended to combat); hence *three*: the meaning the term *actually possesses* (and this meaning is *not* uniquely determined by what happens in observational situations; it contains also theoretical ideas) is still adequate (incorrect result of the application of the second step to the first step); hence, *four*: commonsense is correct, will always be correct, etc. etc. (This train of thought has also given considerable support to the paradigm case argument. Malcolm's paper on Moore and ordinary language proceeds essentially in this fashion). In this connexion we can also see what is to be thought of the argument (which is similar to argument 4 above) that commonsense is essential to science because the scientist can make use of the readings made by his uneducated assistant. First, this argument might show, at the very most, that commonsense *occurs* in the sciences, not that it is *essential* to it. However even that is not established for it has not been shown that the scientist interprets the statement in the same manner in which it is interpreted by his uneducated assistant. As a matter of fact the interpretation used by the assistant is completely irrelevant as is seen from the fact that we still make use of the astronomical data which have been assembled by Babylonian priests in order to prepare them for the appearance of certain *gods* (*their* interpretation, but surely not ours!). Even better, for immediate report the voice of the laboratory assistant might be replaced by the barking of a well conditioned dog who reacts to certain smells and who, while barking, might think "what a lovely sausage."

A final remark. A view such as commonsense is connected with certain expectations, with certain attitudes, and therefore also with a certain way of perceiving the world. This way may become solidified in reaction patterns and add a new level to the more primitive and more animalistic (and animistic) ways of behavior. The attempt to replace commonsense wholly by a different view, if successful, will introduce a new pattern of attitudes, expectations. The world will be "seen" differently. That such a transformation leads to different perceptions is shown by the fact that once a new view is adopted confirming evidence turns up in great number. Before this evidence remained *unnoticed* (and not only unexplained, or unanalysed). Such an attempt at replacement therefore is required for methodological reasons (see (II.) above), and it is also required for making more things "visible." All this means that the new theory will be able to provide its own manifest image. The new level of behavior created by this new manifest image is also necessary for still further levels. Hence, even physiology seems to encourage my view as opposed to weaker views. In this way we may become builders both of new kinds of human beings (more "civilized" ones) and builders of new kinds of commonsense instead of being built, and to this extent dominated, by the commonsense of the past. This view is also the only one that does not pretend that we know nothing when in fact we know a lot and it also takes commonsense more seriously than the other views. It admits that commonsense is a theory, that it may be knowledge and draws from this the consequence that other theories, too, might become commonsense, and should become commonsense if they have been shown to possess certain advantages.

This is all for today. I would be very grateful to have your comments on this. Please let me also know Putnam's reactions. Finally, a very happy Christmas to you.[4]

---

[4] [*Editors' note: The Feyerabend collection at the University of Constance includes a written response by Smart dated December 16, 1963, in which he thanks Feyerabend for the long letter: "No one is so able to stick up for Feyerabend so well as Feyerabend himself" (PF 2-7-46).*]

Preparation (Request for a Sabbatical, 1977)

In the past twenty years I have examined details of science, mainly of the quantum theory, more general problems of scientific method, and the role of science in society.

In quantum theory I made contributions to the theory of measurement, to problems of quantum logic, and I also examined Bohr's interpretation of the formalism of the elementary theory. I showed (1) that Bohr's views survive all objections that have been raised against them; (2) that they are not strong enough to eliminate alternatives; (3) that attempts to strengthen them and to demonstrate the uniqueness of the orthodox interpretation (von Neumanns's proof and refinements) are either inherently faulty, or do not achieve their aim. My work on measurement and quantum logic is summarized in

- "On The Quantum Theory of Measurement" *Observation and Interpretation* ed. Koerner, London 1957.
- "Über die Verwendung nicht klassischer Logiken in der Quantentheorie" *Internationales Forschungszentrum für Grundfragen der Wissenschaften* Vol. I 1965 ed. Weingartner.

The main arguments concerning Bohr's interpretation are found in

- "Eine Bemerkung zum Neumannschen Beweis" *Zs. Physik* 1958.
- "Complementarity" *Proc. Arist. Soc.* Suppl. Vol. 1958.
- "Problems of Microphysics" in *Pittsburgh Studies in the Philosophy of Science* Vol. I ed. Colodny Pittsburgh 1962.

- "On a Recent Critique of Complementarity" *Philosophy of Science* 1968/69.

During the discussion of (2) and (3) my attention was drawn to *methodological rules*, which are taken for granted by scientists and on which the arguments depend in a decisive way. The most complete mathematical proofs are of no avail unless they are combined with these rules. To get a better understanding of the situation I traced the history of the rules and of their influence and I examined them independently of the special context in which I had found them.

The most important rule is *Newton's Rule IV*, which forbids consideration of hypotheses inconsistent with highly confirmed theories (hypotheses, facts) and recommends an inductive development of science from facts to simple generalizations to comprehensive theories. The function of the rule in Newton's methodology is discussed in

- "Classical Empiricism" *The Methodological Heritage of Newton* ed. Butts, Oxford 1970.

Its role in contemporary methodology is examined in my review of Ernest Nagel's *Structure of Science*

- "The Structure of Science" *British Journal for the Philosophy of Science* 1966.

I also examined the origin of the rule and its role in Aristotelian tradition. The results are contained in a large MS [*manuscript*], which remains unpublished.[5]

Next I started examining the validity of the rule. I found that it is violated by successful researchers and that it is inconsistent with a basic principle of empiricism *viz.* to test a theory or a hypothesis against as many facts as possible: alternatives to a highly confirmed theory may produce potentially refuting facts for that theory and are therefore necessary for its examination. And as alternatives do not spring to life in full formal splendor but start as vague guesses it follows that a separation of science on the one hand, myth and

---

[5] [*Editors' note: It is possible that Feyerabend was thinking of section [31] [chapter 6.1] of Philosophy of Nature. We are not aware of another existing manuscript on Newton's Rule IV in Aristotle.*]

metaphysics on the other, is bound to diminish the empirical content of science and to make it more dogmatic. This is shown in some detail in

- "Explanation, Reduction and Empiricism" *Minnesota Studies for the Philosophy of Science* Vol. III, 1962.
- "Problems of Empiricism" *Pittsburgh Studies in the Philosophy of Science* Vol. II, 1965.
- "Von der beschränkten Gültigkeit Methodologischer Regeln" *Neue Hefte für Philosophie* Heft 2/3, 1972.

and summarized in

- "Reply to Criticism" *Boston Studies in the Philosophy of Science* Vol. II 1965.

So far I had tried to show that a separation of the *content* of science and non-science would be detrimental to the former. Metaphysical views, ancient myths, the cosmologies inherent in the various religions are not only valuable reservoirs of ideas, they are possible test cases of the status quo and must be preserved and developed in close connection with science rather than abandoned. Occasionally one even discovers that a myth gave a better account of the world than the scientific theories that replaced it, and so one progresses by returning to earlier ideas. All this means that the separation of science and non-science which plays such an important role in the development of modern thought must cease. However, I still assumed that non-scientific views are (A) not on par with science – they must be *changed* until they become more definite and receive empirical content; and I also assumed (B) that there exists a well defined method for dealing with *any* set of ideas and that it was this method that distinguished a rational examination from enterprises of a different kind.

My studies of myth and of the theatre then made me very doubtful of assumption (A). We know now that the inventors of myth possessed detailed factual knowledge in astronomy, botany, zoology, biology, medicine, sociology, theology, that they tested this knowledge in laboratories and "observatories" such as Stonehenge and used it in their daring voyages. The theories they found are of interest even today, for they often provide better means of diagnosis and therapy than the existing medical doctrines. The myths in which such theories are expressed have a twofold function *viz.* (a) to present knowledge and (b) to utilize the knowledge for furthering social and cosmic

222

harmony. In the past anthropology emphasized only the last function: myth functions as a social glue, but it has no cognitive content. In the course of my research I started suspecting that myths have cognitive content as well. Moreover, considering that the mythmakers *created* culture and *advanced* it to a surprising extent, that the rise of science led to some canvas-cleaning which more than once has thrown out the bad as well as the good I suspected that there might be cases where science and myth are in conflict, but the myth is right, and science is not. I suspected that myths are fully-fledged alternatives to science with a content and a method of presentation of their own. Details are published in chapter 4 of *Against Method* London 1975 – with additions Frankfurt 1976, and in chapter 2 of *Einführung in die Naturphilosophie* Vol. I, Braunschweig 1976.

A study of the early theatre and of Aristotle's account of it led to similar results. A cycle such as the *Oresteia* by Aischylos gives an account of social structures, of the inconsistencies inherent in these structures, it uses dramatized argument to reveal the source of the inconsistencies and makes new suggestions. This is why Aristotle called poetry "more philosophical" than history: history tells what happened. Poetry, especially drama, gives *reasons* for what happened and uses special methods for impressing these reasons on the mind (catharsis). The rise of science led to a separation of these functions and so created the split between knowledge and the arts. Having objected to the separation on theoretical grounds I now found also practical reasons for keeping the parts together. I argued for a stylish science and a poetry with factual content. Style in science is not external embroidery that inhibits serious work, it has an influence on the way in which research is conducted and in which the results of research are understood. This influence can be seen most clearly by comparing the style of the earlier quantum theory with the style of von Neumann and his followers. In the first case statements remain vague, but close to reality, argument is philosophical in character, many different assumptions are tried out. In the second case statements are precise, but removed from reality, argument is mathematical in character and the range of criticism is severely restricted. Brecht's studies of the influence of style on cognitive content in the theatre can be applied and leads to fruitful results. All this is explained in

- "On the Improvement of the Sciences and the Arts and the Possible Identity of the Two" *Boston Studies* Vol. III, 1964/65 "Theater und Ideologiekritik" *Festschrift für Simon Moser* Mesenheim 1967

- "Outlines of a Pluralistic Theory of Knowledge and Action" *Planning for Diversity and Choice* ed. S. Anderson, 1968
- "Lets Make more Movies" *The Owl of Minerva* Prentice Hall 1975.

So far I had recommended abolishing the distinction between science and the rest and my reasons had been

(a) that science needs myth as instruments of criticism
(b) that myths and metaphysics are fully fledged alternatives to science with a form and content of their own and that they provide knowledge not contained in, and perhaps even denied by science
(c) that form (or "style" as I called it above) influences the *attitude* (critical; dogmatic) towards the knowledge presented and therefore should be studied
(d) that the combined study of form and content was once the proper domain of the arts so that a critical science was bound to approach the arts.

I still assumed that there was only one way of dealing with science, myth, the arts and I looked for a general methodology covering all these areas. The methodology was pluralistic and counter-inductive both with respect to form and with respect to content (the main argument here is summarized in "Reply to Criticism" and "On the Improvement . . ."), I used *history* in a rather naive way, as providing instances of rule-use and rule-violation while I relied on abstract argument to give the reasons for such use and such violation.

The next step was the discovery of the following *historical law*: given any rule and any definition of progress, there are always circumstances when progress can only be made by breaking the rule.

This law shows that abstract argument, taken by itself, cannot explain why some procedures succeed while others fail. To arrive at such an explanation we need to know the *material* to which the procedures are applied, the *conditions* in which the application takes place as well as the *general tendencies* that further or hinder work in accordance with the procedures. As science progresses new situations arise that have no counterpart in the past. Is it to be expected that what worked in the past will work in the new circumstances also? We build new instruments and so discover new entities leading to new problems. To solve these problems we need further instruments whose nature depends on the entities discovered and therefore on the original instruments. Is it not plausible to assume that the same applies to our *intellectual* instruments, that the use of new ideas and

224

new procedures creates unforeseen situations which must be explored with the help of new methods? Considerations such as these made me proclaim, in a joking way, a no-method methodology with the principle "anything goes" as its only methodological rule: if you want a rule that works, come what may, then this rule will have to be as empty and ridiculous as the rule "anything goes." A researcher must make up his methodology as he goes along just as he must build his instruments and his theories as he goes along. The research leading up to this stage is summarized in

- "Problems of Empiricism, Part II" *Pittsburgh Studies* Vol. IV, 1970.
- "Against Method" *Minnesota Studies in the Philosophy of Science* Vol. IV, 1970.

It is the point of these two essays that there is not a single logical or methodological rule that can be taken for granted, or can be imposed upon science, come what may. Even such apparently "basic" assumptions as the assumption that factual research is always empirical have their exceptions as is shown in

- "Science without Experience" *Journal of Philosophy* 1969.

Any theory of science, however close to history, can at most be regarded as an account of a passing stage. This is argued in my criticism of Kuhn

- "Consolations for the Specialist" *Criticism and the Growth of Knowledge* Lakatos-Musgrave eds. Cambridge 1970.

Now in the past science was praised both for its successes and because it seemed to be the foremost manifestation of reason. The idea of reason, on the other hand, was always tied to a well defined set of rules: science is rational because there are rules to which it conforms, and because these rules can be shown to agree with reason. Reason has power, because it directs one of the most influential forces in society. Philosophy of science restricted itself to the examination of purely logical problems assuming that these problems were fundamental problems of science as well. The results just mentioned show that none of these ideas agrees with reality: science is not rational in the sense presupposed, reason does not have the power it is assumed to have, philosophy of science as it exists today is, with few exceptions, a special branch of logic and has nothing to do with science itself. This aspect of the situation is discussed in

- "Philosophy of Science: A Subject with a Great Past" *Minnesota Studies* Vol. V. 1970.
- "Die Wissenschaftstheorie – eine bisher unbekannte Form des Irrinns?" *Natur und Geschichte* ed. Huebner, Hamburg 1973.
- *Against Method* London 1975, revised and enlarged German edition Frankfurt 1976.

Result: neither content, nor method, nor the rules of reason enable us to separate science from non-science. Any separation that occurs is a *local phenomenon*, it occurs in certain conditions, between certain parts of science and certain parts of non-science, it cannot be used to infer an *essential difference between the things separated.*

Long range plan:

My long range plan is to construct a theory of knowledge that takes this situation into account. It will differ from the usual theories of knowledge in two ways.

   (1) it will be a theory of the sciences as well as of the arts (humanities). Instead of seeing science and the arts as two different domains obeying different standards it will present them as different parts of one and the same enterprise. In the same way physics and biology are today regarded as parts of one and the same enterprise: science.

   (2) It will not contain any abstract rules. Every rule, the rules of logic included, will be tied to a well specified context and a historical account will be given of the context and of the corresponding use of rules. Historical example and analogy and not logic or an abstract methodology decide how a particular problem is to be solved. A procedure is chosen not because it is inherently reasonable, or because it conforms to the demands of logic, but because it succeeded in the past and because present conditions appear similar to the conditions of its last success. There will of course be cases where history cannot advise and where the researcher must invent new rules and try them out as well as he can. The theory of knowledge I have in mind will consist of three parts: (i) history, (ii) heuristics, (iii) theory of man and associated sciences. History provides the material that enables the researcher to learn his craft. Heuristics provide rules of thumb for the use of this material as well as examples. Finally, the researcher must also become clear about the aims of his research. This is provided by a theory of

man. The associated sciences give a general account of the concrete cases he might have to solve and so facilitate his choice of rules.

The theory of knowledge I have in mind has many similarities with the ancient science of *rhetorics*. (i) corresponds to the collection of speeches which the future rhetorician had to study, (ii) corresponds to the various rules abstracted from these speeches and advice how to invent new rules, (iii) corresponds to the various psychological, sociological, legal, etc. observations that aided him in the use of rules as well as to a list of aims which the speaker might choose on different occasions. (The older theories of knowledge, of course, correspond to the "search for the truth," which was always opposed to rhetorics).

The *history* that is to be assembled under (i) will of course have to be very different from the existent histories which are severely deficient. They are deficient because of the tendency to project modern ways of thought and perception into the past. For example, it is assumed that both the Presocratics and their mythbound predecessors had the same experience when looking at nature and at their fellow humans but that only the former described these experiences clearly and unambiguously while the latter surrounded them with embroideries of a poetic nature. Or it is assumed that Primitives see the world as we do, but misinterpret what they see. Again it is assumed by historians of mathematics such as van der Waerden and von Fritz that the Babylonians and the Greeks had the same notion of external reality but that the former tried to approximate it while the latter tried to give a precise account. Van der Waerden in addition assumes that the only alternative that Thales could possibly perceive was the alternative of proof and approximation. Such a history is of no use for the theory of knowledge I have in mind. It is a reflection of traditional methodological rules, it is constituted in accordance with such rules rather than providing material for their criticism.

In my opinion the ideal historical research proceeds as follows. It explains how *individuals* acquire and change their views, what these views are, what particular historical, psychological, social, theological circumstances influenced their decisions and it describes a *movement* by giving an account of all the individuals that constitute it. For example, it describes the "Copernican Revolution" by explaining how each individual participant became a Copernican. Gaps in the explanation are filled not by appeal to what is considered to be "obvious" by the researcher (such as the methodological rules he regards as basic or forms of thought he cannot imagine being without). They are either left open, or filled with the results of concrete historical research.

227

To take an example: most historians of mathematics compare Greek mathematics with its Babylonian ancestor by saying that the former is systematic and looks for proof while the latter is unsystematic and is content with approximative estimate. This assumes that both the Greeks and the Babylonians had the same entities in mind, they both assumed that these entities had definite properties independently of thought and experimentation, but the former wanted to give an exact and systematic account of these properties while the latter were content with estimates *of the very same properties*. The assumption is quite plausible for us, *but it must not be taken for granted*. After all, it is quite possible that the Babylonians took a constructive approach and regarded a circle as defined by the method used for its production, different methods leading to different types of circles. In this case a circle would be like a house, a certain type of circle being good for some purposes, not so good for others. Similarly it must not be taken for granted that reasons that seem decisive for us were decisive for the early followers of Copernicus.[6] The Copernican revolution, after all, involved not only a change of astronomical ideas, it also involved a change of methodology.

It is clear that the ideal just outlined can hardly ever be achieved. In most cases we have to be content with much less. I showed what can still be done in my account of the transition from Homer to the Presocratics. Here I used internal evidence gained from Homer, contemporary art works (late geometric vase paintings), structure of court poetry, religion, word analysis, comments of later authors (critical or favorable) and so arrived at what I think is a fair account of the Homeric cosmology (anthropology) and of the changes effected by the lyrical poets, the Presocratics, and their followers. It is shown that these changes – which constitute what one might call the rise of rationalism in the West – have advantages (few) as well as disadvantages (many) and that almost all so called epistemological problems go back to them. The material and the relevant methodological comments have been published in ch. 17 of *Against Method* as well

---

[6] Researchers have noted with approval that theological or "spiritualistic" reasons gradually disappear from science. For example, the devil was no longer used as a causal agent in the explanation of various forms of illness. It is not sufficient to state this fact as if it did not need an explanation (obviously, many researchers seem to think, this was the reasonable thing to do, so why should one explain it?). We must ask why such explanations which earlier generations had regarded as natural and even necessary gradually disappear from sight. Was this due to the influence of a particular book, or to political considerations or to a new theology that insisted on the separation of the natural and the divine (devilish)? We cannot understand the scientific revolution unless we try to ask these questions by research, and without recourse to preconceived notions.

as in Vol. I of my *Einführung in die Naturphilosophie*. I have also started with an account of the "scientific revolution" in the 16th and 17th centuries.

Eventually the historical part of the theory of knowledge I am trying to develop will cover the period from the Stone Age up to the present. This will take five to six volumes (the work has been commissioned by Vieweg, Braunschweig). Material for the heuristic and the theory of man will be spread throughout the historical account and will be summarized in a separate volume, once the historical work is completed.

Short range plan:

I have finished the first volume (Stone Age to Parmenides with a detailed account of the invention of a scientific empiricism by Xenophanes) and I am now preparing the second volume (rise of mathematics and astronomy in Greece; Pythagoreans; later Presocratics; Sophists; the social arguments contained in tragedy and their relation to the logic of Zeno). I am also preparing a review of the literature dealing with the 16th and 17th century scientific revolution with a first attempt at a coherent account of major tendencies and events: rise of psychology as a discipline separate from the theological understanding of man; new theories of mental illness and the causes of the gradual disappearance of the witchmania (there is as yet no satisfactory explanation of this phenomenon); decline of Aristotelianism (advantages, disadvantages, causes); function of Neoplatonism, Hermeticism, Rosicrucianism, natural magic and the gradual elimination of spiritual entities from astronomy, physics, biology, psychology; different forms of empiricism: crude (early Royal Society), magical (Agrippa von Nettesheim), theoretical-qualitative (Aristotle), theoretical-quantitative (Galileo), theoretical-constrained (Newton), intuitive (Paracelsians) as well as the anti-empiricism of Bacon. All these schools while urging us to pay attention to facts used different methods and obtained different results, which suggests that the idea of the empirical origin of the scientific revolution in the 16th and 17th centuries is a chimera, created by insufficient analysis. The growing separation of theological and physical matters will also have to be examined. A matter that has been completely neglected is the change of perception (of man; of the heavens) that accompanied the rise of modern science. Such changes start already in the later Middle Ages with the work of the brothers Limburg and they have their first effects in medicine and in medical psychology. They can be traced in the arts, especially in the areas of close collaboration between art and scientific illustration, in

229

the new theories of the arts that are being developed in the 13th to 17th centuries, in critical treatises and descriptions. In tying all these features together I shall try to avoid the following errors:

(1) Easy syntheses: a uniform effect is often due to the fortuitous collaboration of independent causes and so my principle will be to keep tendencies separate as long as the evidence permits.

(2) Enforced rationalism: results which seem reasonable according to later standards have often been achieved on the basis of standards of a very different kind combined with social tendencies which did not permit the standards to act unimpeded. So, my principle will be to prefer unusual and incoherent explanations to plausible and coherent ones as long as the evidence permits.

(3) Transfer of categories: categories which are taken for granted today are often projected into the past and so historical research is replaced by judgments as to what is "obvious" and what is not. For example, it is assumed that the early Greeks and the Babylonians had the same conception of geometrical shapes, but that they formed different theories about them. Almost all research into Pre-Greek mathematics rests on assumptions of this kind (this was pointed out by von Soden in his criticism of Neugebauer). To establish categories one cannot rely on what seems plausible, but has to turn to the evidence. In the case of Babylonian mathematics the evidence is the use of geometrical shapes in art works, the way in which they are referred to in the basic myths, and so on. I have used this method when trying to establish the mental concepts and the concept of man implicit in Homer.

(4) Generalization from individual cases: the actions of an individual scientist can be used to infer general tendencies only if the evidence supports the existence of such tendencies. Individuals with different ideologies often work together and the result is still another ideology. So, my principle will be to enlarge individual differences and to proceed in an "atomistic" way as long as the evidence permits.

(5) Facile sociological hypotheses: intellectual changes are often explained by reference to sociological (historical) tendencies: Puritanism encourages the rise of science, new ideas about the poor and demented (classified together with criminals) give rise to the new class of witches. Such explanations always fail to explain the *ideological* parts of the change: why was *Copernicus* preferred? Why did witches assume support by the *devil*, why did they accept a particular cosmology?

(6) Facile intellectualistic hypotheses: intellectual changes are explained by reference to modern ways of thinking. Example: witchcraft disappeared because of the rise of rationalism. This assumes

230

that the rising rationalism would find neither empirical evidence nor rational reasons favoring witchcraft – which was not the case.

I am also finishing a 200 page account of the theory of knowledge (science) I try to develop, concentrating on the implications for the philosophy of science (and for historical research – see the six points just made). This summary will be published as an introduction to the philosophy of science by Walter de Gruyter, New York; Berlin.[7]

To make my research into the rise of modern science as detailed as possible I am trying to find one or two major figures in the scientific revolution for an individual study of development as described above. I am thinking of Digges, or Robert Recorde, but the final choice will depend on the material available (letters, manuscripts, reports of friends, asides, gossip, reading, etc. etc.). I am convinced that a thorough biographical study including all the tendencies influencing an individual will give a better insight into the factors that determine *scientific change* than either large scale history, or logical reconstruction.

Grant requested

I shall use the summer 76, the summer 77 as well as two sabbatical quarters (spring 77 and fall 77) to carry out the short range plan. I shall spend most of the time in London (Marburg Institute, Wellcome Library, Library of the British Museum), Vienna (Nationalbibliothek, collections of the fine art museum, Volkskundemuseum), Cambridge/ England and Cambridge/Mass (discussions). The Warburg Institute has a large library in the area of interest to me and scholars (Walker, Yates, Gombrich, Schmid), who have advised me in the past and will advise me in the future. The art collections in Vienna have large material for a study of the changes in perception. I shall also examine Purbach and members of the Viennese School as possible subjects for an intellectual biography of the kind described above.

In accordance with this plan I request

(1) a summer supplement for summer 1976 plus round trip San Francisco–Vienna plus funds for xeroxing and typing.
(2) 1/3 supplements for spring and fall 1977 (my sabbatical is at 2/3 salary) plus round trip San Francisco–Vienna plus funds for xeroxing and typing.
(3) a summer supplement for 1977 summer plus roundtrip plus xeroxing-typing.

[7] [*Editors' note: It is not known whether this project was ever realized.*]

Report on 1980 Sabbatical

The *long term project* the sabbatical was supposed to advance is a study of the history of rationalism from antiquity up to the present time and of the influence of rationalism on the sciences, the arts, politics, religion, morality. Rationalism, in this connection, means the use of (a) abstract and observer-independent concepts that can be explained independently of the practice they guide together with (b) a stable logic and methodology. For details cf. the project description for the sabbatical.

The *short term project* consists of two parts *viz.* (a) the history of rationalism in antiquity and the objections that were raised to it and (b) episodes from the history of the arts and the sciences. My work during the sabbatical dealt mainly with the short term project and so, indirectly, with the long term project. It consisted in consultations, the organization of meetings and seminars dealing with the short term projects, the preparation for publication and the publication of papers and discussions from these seminars and individual work and publications of my own.

Persons consulted and subjects discussed: Prof. B. L. von der Waerden (Zürich): ancient mathematics and astronomy; Prof. Burkert (Zürich): transition from Homer to the Presocratics; Prof. Hans Primas (Zürich): reductionism, especially in chemistry; Prof. Gerhard Huber (Zürich): philosophy and the sciences in antiquity; Prof. Walter Hollitscher (Vienna): transition from Homer to the Presocratics, the objectivity of the sciences, science and literature, science and democracy; Prof. Hans Albert (Mannheim/Heidelberg):

232

modern rationalism; Dr. Hans Peter Duerr (Heidelberg/Zürich): paleolithic art; transition from Homer to the Presocratics; Prof. Marx Wartofsky (Boston): scientific trends in the arts – perspective; Prof. T. S. Kuhn (Boston/Berkeley): similarities of revolutionary changes in the sciences and the arts.

Organization of Meetings: In Zürich I organized two sets of meetings, one dealing with the influence of the sciences on traditional subjects such as religion, natural philosophy, the magical view of the world, practical medicine, the other with the role of the sciences in the development of literature, the arts, architecture, etc. The basic idea of the first set of meetings was the following: religious traditions (I invited representatives of the catholic, the Protestant, the Islamic and the Jewish tradition) were thoroughly changed by the rise of the modern sciences. How do the representatives of the traditions now see these changes? Do they regard them as an advantage, as a spring-cleaning, as it were, that removed irrelevant ingredients, or do they think that important elements have been destroyed? The same question was asked of esotericists, creationists, healers. Representatives with different orientations ("progressives," "traditionalists") were invited to explain their views. There was also an interesting discussion of the administrative similarities and differences of church councils and scientific conferences, of the question how the science of theology fits the sciences of today and of the Galileo-trial. The second set of meetings dealt with the complex interactions between the sciences and the arts, for example with the theses, put forth by the metallurgist Cyril Smith that the basic knowledge of materials came from the arts and that philosophical and then scientific theories of matter joined this knowledge only in the 20th century. There was a discussion of Goethe's and Newton's theories of color that emphasized the thoroughly empirical nature of Goethe's research and the dogmatism of Newton.

Publication of Proceedings: the proceedings of the meetings including the papers that were read and part of the discussion were published in two volumes, both edited by Mr. Thomas and myself: *Wissenschaft und Tradition* Zürich 1983 and *Kunst und Wissenschaft* Zürich 1984. I contributed seven papers, the two introductions and discussion remarks.

In addition there are the *following individual publications: Wissenschaft als Kunst*, completed in 1981 but published only 1984. "Ernst Mach's Philosophy of Research and its Relation to Einstein" *Studies in History and Philosophy of Science* 1984 and "Aristotle's Theory of Mathematics and the Continuum" *Midwestern Studies in Philosophy* 1983. The essay deals with the intrusion of science into

233

15th century painting, the paper on Mach introduces a complete reevaluation of Mach's philosophy. Mach is usually regarded as a narrow minded positivist who rejected the theory of relativity because it moved too far away from observation. Actually he suggested such moves and anticipated the rather unusual procedure of Einstein's first paper on relativity. Also Mach was one of the few thinkers to interpret realism as an outgrowth of the tendency to use "objective" and practice-independent concepts and criticized it on these grounds. The Aristotle paper compares the relative superficiality of Galileo's notion of the continuum with Aristotle's much more subtle notion and thus shows that even in the so-called exact sciences the 17th century revolution brought with it gains as well as great losses. This will be an important point both of the long range and the short range project: there is no "clean" progress; there are improvements, but there are also lots of deteriorations. A further paper "Science – Political Party or Instrument of Research?" *Speculations in Science and Technology* 1982 dealt with the question to what extent scientific knowledge is "objective" i.e. separated from the political process and to what extent it must be regarded as part of this process. This question was first discussed by the sophists, Plato and Aristotle and the paper therefore prepares a contemporary reading of these old questions.

Finally, I used the sabbatical to prepare *material for my lectures in Berkeley*. In these lectures, which are on the theory of knowledge and the philosophy of science, I give historical examples of knowledge and knowledge-change from the Stone Age (very inferential, of course, but there exist many conjectures about these matters, for example A. Marshack's magnificent *Roots of Civilization*) via the "Greek Revolution" down to the present century.[8] The examples have to be carefully chosen and the most recent information of them acquired. I use my sabbaticals also to bring examples and presentation up to date.

Berkeley 3/18/85
Paul Feyerabend

---

[8] [*Editors' note: Feyerabend's notes on his Berkeley Lectures from 1975 to 1980 (PF 6-10) at the Constance University Archive also include copies of some text passages and images from Philosophy of Nature (PF 6-10-8).*]

BIBLIOGRAPHY

Ackerman, Phyllis (1960): "Stars and Stories," in: Henry Alexander Murray (ed.), *Myth and Mythmaking*, New York, pp. 90–102.
Aeschylus [*Agamemnon*]: *Agamemnon*, trans. Herbert Weir, Cambridge, MA, 1926.
– [*Libation*]: *Libation Bearers*, trans. Alan H. Sommerstein, in: *Aeschylus II: Oresteia*, Cambridge, MA, pp. 209–354.
– [*Prometheus*]: *Prometheus Bound*, trans. Herbert Weir, Cambridge, MA 1926.
Agrippa, Heinrich Cornelius (1533): *Three Books of Occult Philosophy: A Modern Translation*, trans. Eric Purdue, Iowa City, 2012.
Åkerblom, Kjell (1968): *Astronomy and Navigation in Polynesia and Micronesia: A Survey*, Stockholm.
Althusser, Louis (1965): *For Marx*, trans. Ben Brewster, London, 1968.
Anderson, Edgar (1952): *Plants, Man and Life*, Boston.
Ardrey, Robert (1967): *The Territorial Imperative: A Personal Inquiry into the Animal Origins of Property and Nations*, New York.
Arend, Walter (1933): *Die typischen Scenen bei Homer*, Berlin.
Aquinas, Thomas [*Summa*]: *Summa Theologica*, trans. the Fathers of the English Dominican Province, Chicago, 1952.
Aristotle (1984): *The Complete Works of Aristotle: The Revised Oxford Translation*, ed. Jonathan Barnes, 2 vols., Princeton.
– [*Fragments*]: *Select Fragments*, trans. W. D. Ross, London, 1908–52.
– [*Generation*]: *On Generation and Corruption*, trans. H. H. Joachim, in: Aristotle (1984), pp. 512–54.
– [*Heavens*]: *On the Heavens*, trans. J. L. Stocks, in: Aristotle (1984), pp. 447–511.
– [*Metaphysics*]: *Metaphysics*, trans. Hugh Tredennick, Cambridge, MA, 1989.
– [*Meteorology*]: Meteorology, trans. E. W. Webster, in: Aristotle (1984), pp. 555–625.
– [*Parts*]: *On the Parts of Animals*, trans. W. Ogle, in: Aristotle (1984), pp. 994–1086.
– [*Physics*]: *Physics*, trans. R. P. Hardie and R. K. Gaye, in: Aristotle (1984), pp. 315–446.
– [*Poetics*]: *Poetics*, in: *Aristotle in 23 Volumes. Vol. 23*, trans. W. H. Fyfe, Cambridge, MA/London, 1932.

235

- [*Politics*]: *Politics*, trans. B. Jowett, in: Aristotle (1984), pp. 1986–2129.
- [*Posterior*]: *Posterior Analytics*, trans. G. R. G. Mure, Adelaide, 2007.
- [*Prior*]: *Prior Analytics*, trans. A. J. Jenkinson, in: Aristotle (1984), pp. 39–113.
- [*Prophesying*]: *On Prophesying by Dreams*, trans. J. I. Beare, in: Richard McKeon (ed.), *The Basic Works of Aristotle*, New York, 1966.
- [*Sense*]: *Sense and Sensibilia*, trans. J. I. Beare, in: Aristotle (1984), pp. 693–713.
- [*Sleep*]: *On Sleep and Waking*, trans. David Gallop, in: David Gallop (ed.), *Aristotle on Sleep and Dreams*, 2nd edn., Warminster, 1996.
- [*Soul*]: *On the Soul*, trans. J. A. Smith, in: Aristotle (1984), pp. 641–692.
- [*Topics*]: *Topics*, trans. W. A. Pickard-Cambridge, in: Aristotle (1984), pp. 167–277.
Atkinson, R. J. C. (1960): *Stonehenge*, Harmondsworth.
Atkinson, Richard (1966): "Moonshine on Stonehenge," *Antiquity*, 40, 150, pp. 212–16.
- (1975): "Megalithic Astronomy: A Prehistorian's Comments," *Journal for the History of Astronomy*, 6, pp. 42–52.
Austin, J. L. (1962): *Sense and Sensibilia*, Oxford.
Aveni, Anthony F. (1975): *Archaeoastronomy in Pre-Columbian America*, Austin.
Ayer, Alfred Jules (1940): *The Foundations of Empirical Knowledge*, London.
Bacon, Francis (1605): *The Advancement of Learning*, New York, 1944.
- (1620): *Novum Organum*, trans. and ed. Joseph Devey, New York, 1902.
Beazley, J. D. and Bernard Ashmole (1966): *Greek Sculpture and Painting: To the End of the Hellenistic Period*, Cambridge.
Becher, Johannes R. (1965): "Über Jakob van Hoddis," in: Paul Raabe (ed.), *Expressionismus: Aufzeichnungen und Erinnerungen der Zeitgenossen*, Freiburg, pp. 50–5.
Berkeley, George (1721a): "De Motu: Sive de Motus principio et natura, et de Causa communicationis Motuum," in: Alexander C. Fraser (ed.), *The Works of George Berkeley D. D.; Formerly Bishop of Cloyne. Including his Posthumous Works. In Four Volumes*, Oxford, 1901, *Vol. 1: Philosophical Works*, 1705–21, pp. 501–27.
- (1721b): "An Essay on Motion," trans. Desmond M. Clarke, in: Desmond M. Clarke (ed.), *Berkeley: Philosophical Writings*, Cambridge, 2008, pp. 243–68.
Best, Elsdon (1922): *The Astronomical Knowledge of the Maori, Genuine and Empirical: Including Data Concerning their Systems of Astrogeny, Astrolatry, and Natural Astrology, with Notes on Certain Other Phenomena*, Wellington.
- (1923): *The Maori School of Learning*, Wellington.
Binford, Sally R. (1969): *New Perspectives in Archeology*, Chicago.
Bohm, David (1960): "Commentary," in: Alistair Cameron Crombie (ed.), *Scientific Change*, London, pp. 477–86.
Bohr, Niels (1934): *Atomic Theory and the Description of Nature*, Cambridge.
Boussett, Wilhelm (1960): *Die Himmelsreise der Seele*, Darmstadt.
Brecht, Bertolt (1949a): *Kleines Organon für das Theater*, Frankfurt, 1960.
- (1949b): "A Short Organum for the Theatre," in: John Willett (ed. and trans.), *Brecht on Theatre: The Development of an Aesthetic*, London, 1964, pp. 179–208.
Breuil, Henri (1952): *Four Hundred Centuries of Cave Art*, trans. Mary E. Boyle, Montignac.

236

Budick, Sanford (1970): *Dryden and the Abyss of Light: A Study of "Religio Laici" and the "Hind and the Panther,"* New Haven.

Burkert, Walter (1962): *Weisheit und Wissenschaft: Studien zu Pythagoras, Philolaos und Platon,* Nuremberg.

– (1972): *Lore and Science in Ancient Pythagoreanism,* Cambridge. [*Translator's note: First English edition of Burkert (1962) with extensive revisions of both text and notes.*]

Burkitt, Miles C. (1963): *The Old Stone Age: A Study of Paleolithic Times,* New York.

Calvin, John (1536): "Institutes of the Christian Religion," trans. John T. McNeill, in: John T. McNeill (ed.), *On the Christian Faith: Selections from the Institutes, Commentaries, and Tracts. Edited with an Introduction,* New York, 1956, pp. 3–41.

Cartailhac, Émile (1902): "*Mea culpa* d'un sceptique," *L'Anthropologie,* 13, pp. 348–54.

Castorius and Konrad Miller (1887/88): *Die Peutingersche Tafel.* Reprint of the last rev. edn. by Konrad Miller including his reconstruction of the lost first segment with color rendition of the table as well as brief exposition and 18 cardboard sketches of the traditional Roman travel routes for all countries, Stuttgart, 1962.

Ceram, C. W. (1949a): *Götter, Gräber und Gelehrte: Roman der Archäologie,* Hamburg.

– (1949b): *Gods, Graves and Scholars: The Story of Archeology,* trans. E. B. Garside, New York, 1986.

– (1957): *Götter, Gräber und Gelehrte im Bild,* Hamburg.

Charles, R. H. (ed., 1896): *The Book of the Secret of Enoch,* trans. W. R. Morfill, Oxford.

Childe, V. Gordon (1956): *A Short Introduction to Archeology,* London.

Cohen, I. Bernard (1940): "Roemer and the First Determination of the Velocity of Light," *ISIS,* XXXI, 2 (April), pp. 328–79.

Copernicus, Nicolaus (1543): *On the Revolutions of the Heavenly Spheres,* trans. A. M. Duncan, New York, 1976.

Cornford, Francis M. (1912): *From Religion to Philosophy: A Study in the Origins of Western Speculations,* London, 1957.

– (1937): *Plato's Cosmology: The Timaeus of Plato Translated with a Running Commentary,* London.

– (1950): *The Unwritten Philosophy and Other Essays: Edited with an Introductory Memoir by W. K. C. Guthrie,* Cambridge.

– (1952): *Principium Sapientiae: The Origins of Greek Philosophical Thought,* Cambridge.

Crombie, Alistair Cameron (1953): *Robert Grosseteste and the Origins of Experimental Science, 1100–1700,* Oxford.

Cumont, Franz (1960): *Astrology and Religion among the Greeks and Romans,* New York.

Daniel, Glyn Edmund (1967): *The Origins and Growth of Archeology,* New York.

Descartes, René [*Discourse*]: *Discourse on the Method,* trans. Ian Maclean, Oxford, 2001.

– [*Optics*]: *Optics,* trans. Paul J. Olscamp, in: P. J. Olscamp (ed.), *Discourse on Method, Optics, Geometry, and Meteorology,* Cambridge, 2001, pp. 65–176.

– [*Principles*]: *Principles of Philosophy*, trans. Valentine Rodger Miller and Reese P. Miller, Dordrecht/Boston, 1984.

Dicks, David R. (1966): "Solstices, Equinoxes, and the Presocratics," *Journal of Hellenic Studies*, 86, pp. 26–40.

– (1970): *Early Greek Astronomy to Aristotle*, London.

Dionysius of Halicarnassus [*Thucydides*]: "Thucydides," *Dionysius of Halicarnassus: Critical Essays. Vol. I*, trans. Stephen Usher, Cambridge, MA, 1974.

DK = *Die Fragmente der Vorsokratiker*, Hermann Diels and Walter Kranz (eds.), Berlin 1951. [*Editors' note: The fragments are referenced according to their numbering in this edition. We were not able to verify in each case which translation Feyerabend used or whether he used his own translation.*]

Dodds, Eric (1951): *The Greeks and the Irrational*, Berkeley, 2004.

Dorson, Richard M. (1958): "The Eclipse of Solar Mythology," in: Thomas A. Sebeok (ed.), *Myth: A Symposium*, Bloomington, pp. 15–38.

Duhem, Pierre (1908): *To Save the Phenomena: An Essay on the Idea of Physical Theory from Plato to Galileo*, trans. Edmund Dolan and Chaninah Maschler, Chicago.

Edelstein, Ludwig (1967): *The Idea of Progress in Classical Antiquity*, Baltimore.

Einstein, Albert (1920a): *Äther und Relativitätstheorie: Rede gehalten am 5. Mai 1920 an der Reichs-Universität zu Leiden*, Berlin.

– (1920b): "Ether and the Theory of Relativity: Address Delivered on May 5th 1920 at the University of Leyden," in: Albert Einstein, *Sidelights on Relativity*, London, 1922, pp. 3–24.

Eisler, Robert (1910): *Weltenmantel und Himmelszelt: Religionsgeschichtliche Untersuchungen zur Urgeschichte des antiken Weltbildes*, Munich.

– (1946): *The Royal Art of Astrology*, London.

Emory, Kenneth P. (1974): "The Coming of the Polynesians," *National Geographic*, 146, 6, pp. 732–46.

Engels, Friedrich (1894a): "Herrn Eugen Dürings Umwälzung der Wissenschaft (Anti-Düring)," in: Walter Hollitscher (ed.), *Bücherei des Marxismus-Leninismus*, Berlin, 1953.

– (1894b): *Anti-Dühring: Herr Eugen Dühring's Revolution in Science*, trans. Emile Burns, Moscow, 1947.

Eratosthenes [*Fragments*]: *Erathosthenes' Geography: Fragments Collected and Translated, with Commentary and Additional Material*, ed. and trans. Duane W. Roller, Princeton, 2010.

Evans-Pritchard, Edward E. (1937): *Witchcraft, Oracles and Magic among the Azande*, Oxford.

– (1940): *The Nuer: A Description of the Modes of Livelihood and Political Institutions of a Nilotic People*, Oxford.

– (1964): *Social Anthropology and Other Essays*, New York.

Feyerabend, Paul (1965): "Reply to Criticism: Comments on Smart, Sellars and Putnam," in: R. Cohen and M. Wartofsky (eds.), *Proceedings of the Boston Colloquium for the Philosophy of Science 1962–1964: In Honor of Philip Frank*, Boston Studies in the Philosophy of Science, vol. 2, New York, pp. 223–61. Reprinted in Paul Feyerabend, *Realism, Rationalism and Scientific Method: Philosophical Papers. Vol. 1*, Cambridge, 1985, pp. 104–31.

– (1968): "On the Recent Critique of Complementarity: Part I," *Philosophy of Science*, 35, 4, pp. 309–31.

- (1969a): "On the Recent Critique of Complementarity: Part II," *Philosophy of Science*, 36, 1, pp. 82–105.
- (1969b): "Science without Experience," *Journal for Philosophy*, 66, pp. 791–4.
- (1972a): "Von der beschränkten Gültigkeit methodologischer Regeln," *Neue Hefte für Philosophie*, 2/3, pp. 124–71.
- (1972b): "On the Limited Validity of Methodological Rules," in: Paul Feyerabend, *Knowledge, Science and Relativism*, ed. John Preston, Cambridge, 2008, pp. 138–80.
- (1975): *Against Method: Outline of an Anarchistic Theory of Knowledge*, London/New York, 1993.
- (1978): *Science in a Free Society*, New York, 1982.
- (1983a): "Science as Art: A Discussion of Riegl's Theory of Art and an Attempt to Apply it to the Sciences," *Art and Text*, 12–13 (Summer 1983–Autumn 1984), pp. 16–46.
- (1983b): *Wissenschaft als Kunst*, Frankfurt, 1984.
- (1984): "Xenophanes: A Forerunner of Critical Rationalism?," trans. John Krois, in: Gunnar Anderson (ed.), *Rationality in Science and Politics*, Boston Studies in the Philosophy of Science, vol. 79, Dordrecht/Boston/Lancaster, 1984, pp. 95–109.
- (1987): *Farewell to Reason*, New York.
- (1994): *Killing Time: The Autobiography of Paul Feyerabend*, Chicago, 1995.
- (1999): *Conquest of Abundance*, Chicago.
Finley, Moses I. (1970): *Early Greece: The Bronze and Archaic Ages*, London.
Forsdyke, Edgar John (1964): *Greece Before Homer: Ancient Chronology and Mythology*, New York.
Fränkel, Herrmann (1960): *Wege und Formen frühgriechischen Denkens: Literarische und philosophiegeschichtliche Studien*, Munich, 2nd rev. edn.
Frankfort, Henri, H. A. Gronewegen-Frankfort, John A. Wilson, et al. (eds., 1949): *Before Philosophy: The Intellectual Adventure of Ancient Man. An Essay on Speculative Thought in the Ancient Near East*, Harmondsworth.
Freeman, Kathleen (1948): *Ancilla to the Pre-Socratic Philosophers*, Harvard, 1957.
Freud, Sigmund (1900a): "Die Traumdeutung," in: Alexander Mitscherlich, Angela Richards, and James Strachey (eds.), *Sigmund Freud Studienausgabe*, Frankfurt, 1972, vol. 2, pp. 21–588.
- (1900b): *The Interpretation of Dreams*, trans. A. A. Brill, New York, 1913.
Friedländer, Paul (1954): *Platon. Band I: Seinswahrheit und Lebenswirklichkeit*, Berlin.
Fritz, Kurt von (1938): *Philosophie und sprachlicher Ausdruck bei Demokrit, Plato und Aristoteles*, Leipzig.
- (1946): "NOUS, NOEIN, and their Derivatives in Pre-Socratic Philosophy (Excluding Anaxagoras)," *Classical Philology*, 41, pp. 12–34.
- (1962): *Antike und moderne Tragödie: Neun Abhandlungen*, Berlin.
- (1967): *Die griechische Geschichtsschreibung*, Berlin.
- (1971): *Grundprobleme der Geschichte der antiken Wissenschaft*, Berlin/New York.
Galilei, Galileo [*Motion*]: *On Motion*, trans. I. E. Drabkin, in: I. E. Drabkin and Stillman Drake (eds.), *On Motion and On Mechanics*, Madison, 1960, pp. 13–114.

Gay, Peter (1970): *The Enlightenment: An Interpretation. Vol. 2: The Science of Freedom*, London.

Giedion, Siegfried (1962): *The Beginnings of Art*, Oxford.

– (1964): *Die Entstehung der Kunst*, Cologne.

Gigon, Olof (1968): *Der Ursprung der griechischen Philosophie: Von Hesiod bis Parmenides*, Basel, 2nd edn.

Godelier, Maurice (1971): "Myth and History," in: *New Left Review*, 69, pp. 93–112.

Golson, Jack (1972): *Polynesian Navigation: A Symposium on Andrew Sharpe's Theory of Accidental Voyages*, Wellington.

Gombrich, Ernst H. (1960): *Art and Illusion: A Study in the Psychology of Pictorial Representation*, Washington.

Goodall, Jane (1971): *In the Shadow of Man*, London.

Goodenough, W. H. (1953): *Native Astronomy in the Central Carolines*, Philadelphia.

Graziosi, Paolo (1960): *Paleolithic Art*, London.

Griaule, Marcel (1965): *Conversations with Ogotemmêli: An Introduction to Dogon Religious Ideas*, Oxford.

Grimble, Arthur (1931): "Gilbertese Astronomy and Astronomical Observations," *Journal of Polynesian Society*, 40, 160, pp. 197–224.

Groenewegen-Frankfort, Henriette A. (1951): *Arrest and Movement: An Essay on Space and Time in the Representational Art of the Ancient Near East*, London.

Grube, Georges M. A. (1965): *The Greek and Roman Critics*, London.

Guthrie, William K. C. (1951): *The Greeks and Their Gods*, Boston.

– (1962): *A History of Greek Philosophy. I: The Early Presocratics and the Pythagoreans*, Cambridge.

– (1965): *A History of Greek Philosophy. II: The Presocratic Tradition from Parmenides to Democritus*, Cambridge.

Hampe, Roland (1952): *Die Gleichnisse Homers und die Bildkunst seiner Zeit*, Tübingen.

Hanfmann, G. M. S. (1957): "Narration in Greek Art," *American Journal of Archeology*, 61, pp. 71–8.

Hart, Heinrich (1907): "Literarische Erinnerungen," *Heinrich Hart: Gesammelte Werke. Vol. 3*, ed. Julius Hart, Berlin.

Hauser, Arnold (1951): *A Social History of Art*, New York.

Hawkes, Jacquetta (1967): "God in the Machine," *Antiquity*, 41, 163, pp. 174–80.

Hawkins, Gerald S. (1965): *Stonehenge Decoded*, Garden City.

– (1968): "Astro-Archaeology," *Vistas in Astronomy*, 10, pp. 45–54.

Hegel, Georg Wilhelm Friedrich (1801a): "Differenz des Fichte'schen und des Schelling'schen Systems der Philosophie," in: Hartmut Buchner and Otto Pöggeler (eds.), *Georg Wilhelm Friedrich Hegel: Gesammelte Werke*, vol. 4, Hamburg, 1968, pp. 1–92.

– (1801b): *Difference Between Fichte's and Schelling's System of Philosophy*, trans. W. Cerf and H. S. Harris, Albany, 1977.

– (1802a): "Verhältniss des Sceptizismus zur Philosophie, Darstellung seiner verschiedenen Modifikationen, und Vergleichung des Neuesten mit dem Alten," in: Hartmut Buchner and Otto Pöggeler (eds.), *Georg Wilhelm Friedrich Hegel: Gesammelte Werke*, vol. 4, Hamburg, 1968, pp. 197–238.

- (1802b): "Relationship of Skepticism to Philosophy, Exposition of Its Different Modifications and Comparison to the Latest Form with the Ancient One," trans. H. S. Harris, in: George di Giovanni and H. S. Harris (eds.), *Between Kant and Hegel*, Albany, 1985, pp. 311–62.
- (1804/5a): "Jenenser Logik," in: Georg Lasson (ed.), *Georg Wilhelm Friedrich Hegel: Jenenser Logik, Metaphysik und Naturphilosophie*, Hamburg, 1967, pp. 1–129.
- (1804/5b): "Logic," trans. George di Giovanni, in: John W. Burbidge and George di Giovanni (eds.), *The Jena System, 1804/5: Logic and Metaphysics*, Kingston and Montreal, 1986, pp. 3–130.
- (1812a): *Wissenschaft der Logik I: Die objektive Logik*, in: Georg Lasson (ed.), *Georg Wilhelm Friedrich Hegel: Wissenschaft der Logik. Erster Teil*, Hamburg, 1963.
- (1812b): *The Science of Logic. Vol. 1: The Objective Logic. Book I: The Doctrine of Being*, in: G. W. F. Hegel, *The Science of Logic*, trans. George di Giovanni, Cambridge, 2010, pp. 45–336.
- (1813a): *Wissenschaft der Logik II: Die Lehre vom Wesen*, in: Georg Lasson (ed.), *Georg Wilhelm Friedrich Hegel: Wissenschaft der Logik. Zweiter Teil*, Hamburg, 1963.
- (1813b): *The Science of Logic. Vol. 1: The Objective Logic. Book II: The Doctrine of Essence*, in: G. W. F. Hegel, *The Science of Logic*, trans. George di Giovanni, Cambridge, 2010, pp. 337–506.
- (1830a): *Enzyclopädie der philosophischen Wissenschaften im Grundriss: Auf der Grundlage der Lassonschen Ausgabe (Hegels Sämtliche Werke: Kritische Ausgabe. Vol. 5)*, ed. Johannes Hoffmeister, Leipzig, 1930.
- (1830b): *The Encyclopaedia Logic: Part 1 of the Encyclopaedia of Philosophical Sciences*, trans. T. F. Geraets, W. A. Suchting, and H. S. Harris, Indianapolis, 1991.
- (1830c): *Philosophy of Nature: Part Two of the Encyclopaedia of Philosophical Sciences*, trans. Michael John Petry, 3 vols., London, 1970.
- (1830d): *Hegel's Philosophy of Mind: Being Part Three of the Encyclopedia of the Philosophical Sciences*, trans. William Wallace, Oxford, 1971.
Heidel, Alexander (1949): *The Gilgamesh Epic and Old Testament Parallels*, Chicago.
Henning, Richard (1936): *Terrae Incognitae: Eine Zusammenstellung und kritische Bewertung der wichtigsten vorkolumbischen Entdeckungsreisen an Hand der darüber vorliegenden Originalberichte. I: Altertum bis Ptolemäus*, Leiden.
Heraclitus of Ephesus [*Fragments*]: *The Fragments of Heraclitus*, trans. G. W. T. Patrick, Baltimore, 1889.
Herder, Johann Gottfried (1769a): "Journal meiner Reise im Jahre 1769," in: Karl Kerényi (ed.), *Die Eröffnung des Zugangs zum Mythos*, Darmstadt, 1976, pp. 4–5.
- (1769b): *Journal of my Voyage in the Year 1769*, trans. Frederick M. Barnard, in: F. M. Barnard (ed.), *Herder on Social and Political Culture*, Cambridge, 1969, pp. 63–178.
Herodotus [*Histories*]: *Histories*, trans. A. D. Godley, Cambridge, 1920.
Hesiod [*Erga*]: *Works and Days*, trans. H. G. Evelyn White, Harvard, 1914.
- [*Theogony*]: *Theogony*, in: *Hesiod, the Homeric Hymns, and Homerica*, trans. H. G. Evelyn White, Harvard, 1914, pp. 78–153.

Homer [*Iliad*]: *The Iliad*, trans. A. T. Murray, Cambridge, MA/London, 1924.
– [*Odyssey*]: *The Odyssey*, trans. A. T. Murray, Cambridge, MA/London, 1919.
Hopkins, Matthew (1640): *The Discovery of Witches: A Study of Master Matthew Hopkins, commonly call'd Witch Finder Generall*, ed. Montague Summers, London, 1939.
Hoyle, Fred (1966): "Speculations on Stonehenge," *Antiquity*, 40, 160, pp. 262–76.
– (1972): *From Stonehenge to Modern Cosmology*, San Francisco.
Hughes, H. Stuart (1964): *History as Art and as Science: Twin Vistas on the Past*, New York.
Jacobsen, Thorkild (1949): "The Cosmos as a State," in: Henri Frankfort, H. A. Groenewegen-Frankfort, John A. Wilson, et al. (eds.), *Before Philosophy: The Intellectual Adventure of Ancient Man. An Essay on Speculative Thought in the Ancient Near East*, Harmondsworth, pp. 125–84.
Jaeger, Werner (1923a): *Aristoteles: Grundlegung einer Geschichte seiner Entwicklung*, Berlin.
– (1923b): *Aristotle: Fundamentals of the History of his Development*, trans. Richard Robinson, Oxford, 1948.
Jeremias, Alfred (1929): *Handbuch der altorientalischen Geisteskultur*, Berlin/Leipzig, 2nd rev. edn.
Jung, Carl Gustav (1912a): *Wandlungen und Symbole der Libido: Beiträge zur Entwicklungsgeschichte des Denkens*, Leipzig.
– (1912b): *Psychology of the Unconscious: A Study of the Transformations and Symbolisms of the Libido. A Contribution to the History of the Evolution of Thought*, trans. Beatrice M. Hinkle, New York, 1916.
– (1916): "Concerning the Two Kinds of Thinking," in: Jung (1912b: 9–36).
– (1944a): *Psychologie und Alchemie*, Zurich.
– (1944b): *Psychology and Alchemy*, trans. Gerhard Adler and R. F. C. Hull, in: *Collected Works of C. G. Jung. Vol. 12*, Princeton, 1980, 2nd edn.
Jung, C. G. and Karl Kerényi (1951a): *Einführung in das Wesen der Mythologie*, Zurich.
– (1951b): *Introduction to a Science of Mythology*, London.
Kahn, Charles H. (1960): *Anaximander and the Origins of Greek Cosmology*, New York.
Kearney, Hugh (1967): *Origins of the Scientific Revolution*, New York.
Kearney, Hugh and Kurt Neff (1971): *Und es entstand ein neues Weltbild: Die wissenschaftliche Revolution vor einem halben Jahrtausend*, Munich.
Kerényi, Karl (1967): *Die Eröffnung des Zugangs zum Mythos, ein Lesebuch*, Darmstadt.
Kierkegaard, Søren (1846): *Concluding Unscientific Postscript to Philosophical Fragments*, trans. and ed. David F. Swenson and Walter Lowrie, Princeton, 1941.
Kirk, Geoffrey S. (ed., 1964): *The Language and Background of Homer: Some Recent Studies and Controversies*, Views and Controversies, Cambridge.
– (1965): *Homer and the Epic: A Shortened Version of "The Songs of Homer,"* Cambridge.
– (1970): *Myth: Its Meaning and Functions in Ancient and Other Cultures*, Cambridge.
Kirk, Geoffrey S. and John E. Raven (1957): *The Presocratic Philosophers: A Critical History with a Selection of Texts*, Cambridge.

König, Marie E. P. (1954): *Das Weltbild des eiszeitlichen Menschen*, Marburg.
- (1966): "Die Symbolik des urgeschichtlichen Menschen," *Symbolon*, 5, pp. 121–61.
- (1973): *Am Anfang der Kultur: Die Zeichensprache des frühen Menschen*, Berlin.
Körner, Otto (1929): *Die ärztlichen Kenntnisse in Ilias und Odyssée*, Munich.
- (1930): *Die Homerische Tierwelt*, Munich, 2nd edn. rev. and amended for zoologists and philologists.
Koyré, Alexandre (1957): *From the Closed World to the Infinite Universe*, Baltimore, 1986.
Krafft, Fritz (1971): *Geschichte der Naturwissenschaft*, Freiburg.
Krämer, Augustin (1903a): *Die Samoa-Inseln: Entwurf einer Monographie mit besonderer Berücksichtigung Deutsch-Samoas*, Stuttgart.
- (1903b): *The Samoa Islands: An Outline of a Monograph with Particular Consideration of German Samoa*, trans. Theodore Verhaaren, Honolulu, 1994–5.
Kühn, Herbert (1965): *Eiszeitkunst: Die Geschichte ihrer Erforschung*, Göttingen.
Kühner, Raphael, Friedrich Blass, and Bernhard Gerth (1966): *Ausführliche Grammatik der griechischen Sprache*, Darmstadt.
Kurz, Gebhard (1966): *Darstellungsformen menschlicher Bewegung in der Ilias*, Heidelberg.
Laing, Ronald D. (1967): *The Politics of Experience and the Bird of Paradise*, London.
Lakatos, Imre (1963/4): "Proofs and Refutations," *British Journal for the Philosophy of Science*, 14, pp. 1–25, 120–39, 221–45, 296–342.
- (1970): "Falsification and the Methodology of Scientific Research Programmes," in: Lakatos and Musgrave (1970: 91–196).
Lakatos, Imre and Alan Musgrave (eds., 1970): *Criticism and the Growth of Knowledge*, Cambridge.
Laming-Emperaire, Annette (1959): *Lascaux. Paintings and Engravings*, Harmondsworth.
- (1962): *La Signification de l'Art Rupestre Paléolithique: Méthodes et Applications*, Paris.
Lattimore, Richmond A. (trans., 1951): *The Iliad of Homer*, Chicago.
Leibniz, Gottfried Wilhelm (1715–16): "From the Letters to Clarke (1715–16)," in: Roger Ariew (ed.), *Leibniz: Philosophical Essays*, Cambridge, MA, 1989, pp. 320–46.
- [ad Cartesii]: "Animadversiones ad Cartesii principia philosophiae," in: Gottschalk E. Guhrauer (ed.), *Leibnitz's Animadversiones ad Cartesii principia philosophiae: Aus einer noch ungedruckten Handschrift mitgetheilt*, Bonn, 1844, pp. 1–80.
Lenin, Vladimir Ilyich (1929): "Conspectus of Hegel's Book *The Science of Logic*," in: *Lenin's Collected Works. Vol. 38*, 4th edn., trans. Clemence Dutt, Moscow, 1976, pp. 85–241.
Leroi-Gourhan, André (1967): *Treasures of Prehistoric Art*, trans. Norbert Guterman, New York.
- (1968): "The Evolution of Paleolithic Art," *Scientific American*, 218, pp. 58–70.
Levinson, Michael, Ralph Gerard Ward, John Winter Webb, et al. (1973): *The Settlement of Polynesia: A Computer Simulation*, Minneapolis.

Lévi-Strauss, Claude (1955): *A World on the Wane*, trans. John Russell, New York, 1963.
- (1958): *Structural Anthropology*, trans. Claire Jacobson and Brooke Grundfest Schoepf, Chicago, 1963.
- (1962a): *The Savage Mind*, trans. John Weightman and Doreen Weightman, Chicago, 1966.
- (1962b): *Totemism*, trans. Rodney Needham, Boston, 1963.
- (1964): *The Raw and the Cooked: Introduction to a Science of Mythology*, trans. John Weightman and Doreen Weightman, Chicago, 1969.
Lewis, David (1966a): "An Experiment in Polynesian Navigation," *Journal of the Institute of Navigation*, 19, pp. 154–68.
- (1966b): "Stars of the Sea Road," *Journal of the Polynesian Society*, 75, 1, pp. 85–94.
- (1971): "A Return Voyage Between Puluwat and Saipan Using Micronesian Navigational Techniques," *Journal of the Polynesian Society*, 80, pp. 437–48.
- (1974a): "Voyaging Stars. Aspects of Polynesian and Micronesian Astronomy," in: F. R. Hodson (ed.), *The Place of Astronomy in the Ancient World*, London, pp. 133–48.
- (1974b): "Wind, Wave, Star, and Bird," *National Geographic*, 146, 6, pp. 747–55.
Lobeck, Christian August (1829): *Aglaophamus sive de theologiae mysticae Graecorum causis libri tres*, Königsberg.
Lockyer, Joseph Norman (1964): *The Dawn of Astronomy: A Study of the Temple Worship and Mythology of the Ancient Egyptians*, Cambridge.
Loewy, Emanuel (1900a): *Die Naturwiedergabe in der älteren griechischen Kunst*, Rome.
- (1900b): *The Rendering of Nature in Early Greek Art*, London, 1907.
Lorenz, Konrad (1935a): "Der Kumpan in der Umwelt des Vogels," in: K. Lorenz (ed.), *Über tierisches und menschliches Verhalten: Aus dem Werdegang der Verhaltenslehre. Gesammelte Abhandlungen*, vol. 1, Munich, pp. 115–282.
- (1935b): "The Companion in the Bird's World," *Auk*, 54, 1 (1937), pp. 245–73.
Lowie, Robert Harry (1937): *The History of Ethnological Theory*, New York.
Luther, Martin (1520): "A Prelude by Martin Luther on the Babylonian Captivity of the Church," trans. Albert T. W. Steinhauser, in: *Works of Martin Luther with Introductions and Notes*, Philadelphia, 1915, pp. 167–293.
MacKie, E. W. (1974): "Archeological Tests on Supposed Prehistoric Astronomical Sites in Scotland," in: Frank Roy Hudson and D. G. Kendall (eds.), *The Place of Astronomy in the Ancient World: A Joint Symposium of the Royal Society and the British Academy*, London, pp. 164–94.
Mailer, Norman (1970): *Of a Fire on the Moon*, Boston.
Makemson, Maud Worcester (1941): *The Morning Star Rises: An Account of Polynesian Astronomy*, London.
Maranda, Pierre (1972): *Mythology: Selected Readings*, Harmondsworth.
Marshack, Alexander (1971): *The Roots of Civilization: The Cognitive Beginnings of Man's First Art, Symbol, and Notation*, New York.
- (1972a): "Cognitive Aspects of Upper Paleolithic Engravings," *Current Anthropology*, 13, pp. 445–61.
- (1972b): "Upper Paleolithic Notation and Symbol: Sequential Microscopic

Analyses of Magdalenian Engravings Document Possible Cognitive Origins of Writing," *Science*, 178, pp. 817–28.

– (1975): "Exploring the Mind of Ice Age Man," *National Geographic*, 147, 1, pp. 64–89.

Matz, Friedrich (1950): *Geschichte der griechischen Kunst*, vol. 1: *Die geometrische und die früharchaische Form*, Frankfurt.

McLuhan, Marshall (1965): *The Gutenberg Galaxy: The Making of Typographic Man*, Toronto.

Meissner, Bruno (1925): *Babylonien und Assyrien*, 2 vols., Heidelberg.

Merton, Robert K. (1938): *Science, Technology and Society in Seventeenth Century England*, New York, 1970.

Misner, Charles W., Kip S. Thorne, and John Archibald Wheeler (1973): *Gravitation*, San Francisco.

More, Louis Trenchard (1962): *Isaac Newton: A Biography*, New York.

Murray, Gilbert (1934): *The Rise of the Greek Epic*, London, 4th edn.

Murray, Henry Alexander (ed., 1960): *Myth and Mythmaking*, New York.

Nestle, Wilhelm (1942): *Vom Mythos zum Logos: Die Selbstentfaltung des griechischen Denkens von Homer bis auf die Sophistik und Sokrates*, Stuttgart, 2nd edn.

Neugebauer, Otto (1952): "Tamil Astronomy: A Study in the History of Astronomy in India," *Osiris*, 10, pp. 252–76.

Neuwaldt, Hermann, Wilhelm Adolf Scribonius, and Heinrich Julius (1584): *Exegesis Purgationis Sive Examinis Sagarum super aquam frigidam proiectarum: In qua Refutata opinione Geuilhelmi Adolphi Scribonij, de huius purgationis & aliarum similium origine, natura, & veritate agitur; omnibus ad rerum gubernacula sedentibus maxime necessaria. Autore Hermanno Neuwaldt, Medicine Doctore*, Helmstedt.

Newton, Isaac (1713): *Newton's Principia: The Mathematical Principles of Natural Philosophy*, trans. Andrew Motte, New York, 1846.

Nietzsche, Friedrich (1873a): "Über Wahrheit und Lüge im außermoralischen Sinn," in: Karl Schlechta (ed.), *Friedrich Nietzsche: Werke in drei Bänden*, Munich, 1956, vol. 3, pp. 309–22.

– (1873b): "On Truth and Lies in a Nonmoral Sense," trans. Walter Kaufmann, in: W. Kaufmann (ed.), *The Portable Nietzsche*, London, 1977, pp. 42–6.

Nilsson, Martin Persson (1932): *The Mycenaean Origin of Greek Mythology*, Berkeley.

– (1940): *Greek Popular Religion*, New York.

Otto, Walter Friedrich (1947a): *Die Götter Griechenlands: Das Bild des Göttlichen im Spiegel des griechischen Geistes*, Frankfurt.

– (1947b): *The Homeric Gods: The Spiritual Significance of Greek Religion*, trans. M. Hadas, 1979.

– (1956): "Der ursprüngliche Mythos," in: Karl Kerényi (ed.), *Die Eröffnung des Zugangs zum Mythos*, Darmstadt, pp. 271–8.

Page, Denys L. (1959): *History and the Homeric Iliad*, Berkeley/Los Angeles.

Parry, Adam (1964): "The Language of Achilles," in: Geoffrey S. Kirk (ed.), *The Language and Background of Homer: Some Recent Studies and Controversies*, Cambridge, pp. 48–54.

Parry, Milman (1930): "Studies in the Epic Technique of Oral Verse-Making. I: Homer and Homeric Style," *Harvard Studies in Classical Philology*, 41, pp. 73–148.

Pearson, Lionel I. C. (1939): *Early Ionian Historians*, Oxford.

Pfeiffer, Rudolf (1968): *History of Classical Scholarship from the Beginnings to the End of the Hellenistic Age*, Oxford.

Pfuhl, Ernst (1923): *Malerei und Zeichnung der Griechen*, Munich.

Piaget, Jean (1954): *The Construction of Reality in the Child*, trans. Margaret Cook, New York.

Pindar [*Olympian*]: *Olympian and Pythian Odes: With an Introductory Essay, Notes, and Indexes*, trans. and ed. Basil Gildersleeve, Cambridge, 2010.

Plato [*Euthyphro*]: *Euthyphro*, trans. Harold North Fowler, Cambridge, 1966.

– [*Laws*]: *Laws*, trans. R. G. Bury, Harvard, 1926.

– [*Phaedo*]: *Phaedo*, trans. Harold North Fowler, Cambridge, 1966.

– [*Phaedrus*]: *Phaedrus*, trans. Harold North Fowler, Cambridge, 1925.

– [*Republic*]: *Republic*, trans. Paul Shorey, Harvard, 1935.

– [*Timaeus*]: *Timaeus*, trans. Robert Gregg Bury, Cambridge, 1929.

Plutarch [*Dinner*]: *The Dinner of the Seven Wise Men*, trans. Frank C. Babbitt, in: T. E. Page et al. (eds.), *Plutarch's Moralia. Vol. II (86b–171f.)*, Cambridge, MA, 1956, pp. 346–449.

– [*Face*]: *Concerning the Face which Appears in the Orb of the Moon*, trans. Harald Cherniss, in: T. E. Page et al. (eds.), *Plutarch's Moralia. Vol. XII (920a–999b)*, Cambridge, MA, 1957, pp. 1–223.

Pólya, George (1954): *Mathematics and Plausible Reasoning*, 2 vols., Princeton.

Popper, Karl (1945): *The Open Society and Its Enemies. Vol. I: The Spell of Plato*, London.

– (1958): "Back to the Presocratics," in: Karl Popper, *Conjectures and Refutations: The Growth of Scientific Knowledge*, London, 1998, pp. 136–65.

– (1967): "Quantum Mechanics without 'the Observer,'" in: Mario Bunge (ed.), *Quantum Theory and Reality*, Berlin/Heidelberg/New York, pp. 7–44.

Proclus [*Commentary*]: *Commentary on Plato's Timaeus*, ed. and trans. Dirk Baltzly, Cambridge, 2011.

Rank, Otto (1909a): *Der Mythus von der Geburt des Helden: Versuch einer psychologischen Mythendeutung*, Leipzig/Vienna.

– (1909b): *The Myth of the Birth of the Hero: A Psychological Interpretation of Mythology*, trans. F. Robbins, New York, 1959.

Reinhardt, Karl (1959): *Parmenides und die Geschichte der griechischen Philosophie*, Frankfurt, 2nd edn.

Renfrew, Colin (1971): "Carbon 14 and the Prehistory of Europe," *Scientific American*, 10, pp. 63–72.

– (1973): *Before Civilization: The Radiocarbon Revolution and Prehistoric Europe*, London.

Rosenfeld, Léon (1967): "Niels Bohr in the Thirties," in: Stefan Rozenthal (ed.), *Niels Bohr: His Life and Work as Seen by His Friends and Colleagues*, Amsterdam/New York, pp. 114–36.

Roth, Alfred G. (1945): *Die Gestirne in der Landschaftsmalerei des Abendlandes: Ein Beitrag zum Problem der Natur in der Kunst*, Bern.

Rousseau, Jean-Jacques [*Reply*]: "Last Reply," in: Victor Gourevitch (ed.), *Rousseau: The Discourses and Other Early Political Writings*, Cambridge, 1997, pp. 63–85.

Santillana, Giorgio de and Hertha von Dechend (1969): *Hamlet's Mill: An Essay on Myth and the Frame of Time*, Boston.

Sarton, George (1947): *Introduction to the History of Science III. I: Science and Learning in the Fourteenth Century*, Washington.

– (1959): *A History of Science. I: Ancient Science Through the Golden Age of Greece*, Cambridge.

Schachermeyr, Fritz (1966): *Die frühe Klassik der Griechen*, Stuttgart.

Schäfer, Heinrich (1963): *Von ägyptischer Kunst: Eine Grundlage*, Wiesbaden, 4th rev. edn, ed. Emma Brunner-Traut with postscript.

Schmitt, Charles B. (1969): "Experience and Experiment: A Comparison of Zabarella's View with Galileo's in *De Motu*," *Studies in the Renaissance*, 16, pp. 80–138.

Schramm, Matthias (1963): *Ibn al-Haythams Weg zur Physik*, Wiesbaden.

Schwabl, Hans (1953): "Sein und Doxa bei Parmenides," *Wiener Studien*, 66, pp. 50–75.

– (1958): "Weltschöpfung," in: A. Pauly and G. Wissowa (eds.), *Realencyclopädie der classischen Altertumswissenschaft*, vol. IX, pp. 1433–518.

– (1964): "Anaximander: Zu den Quellen und seiner Einordnung im vorsokratischen Denken," *Archiv für Begriffsgeschichte*, 9, pp. 59–72.

– (1965): "Charles H. Kahn: Anaximander and the Origins of Greek Cosmology," *Gnomon*, 37, pp. 225–8.

Scribonius, Wilhelm Adolf (1583): *De examine et purgatione sagarum per aquam frigidam: epistola*, Lemgo.

Sebeok, Thomas A. (1965): *Myth: A Symposium*, Bloomington.

Sextus Empiricus [*Pyrrhonism*]: *Outlines of Pyrrhonism*, trans. and ed. Benson Mates, Oxford, 1996.

– [*Logicians*]: *Against the Logicians*, trans. and ed. Richard Bett, Cambridge, 2005.

Sharp, Andrew (1963): *Ancient Voyagers in Polynesia*, Sydney.

Simplicius [*Physics*]: *On Aristotle's Physics 1.3–4*, Ancient Commentators on Aristotle, trans. C. C. W. Taylor and Pamela M. Huby, Bristol, 2011.

Singer, Charles, E. J. Holmyard, and A. R. Hall (1967): *A History of Technology. Vol. I: From Early Times to Fall of Ancient Empires*, Oxford.

Smith Bowen, Eleonore (1954): *Return to Laughter*, London.

Snell, Bruno (1924): *Die Ausdrücke für den Begriff des Wissens in der vorplatonischen Philosophie*, Berlin.

– (1946a): *Die Entdeckung des Geistes: Studien zur Entstehung des europäischen Denkens bei den Griechen*, Hamburg, 2nd amended edn.

– (1946b): *The Discovery of the Mind: The Greek Origins of European Thought*, trans. T. G. Rosenmeyer, New York, 1960. [*Editors' note: Feyerabend most likely used this English translation rather than the German one.*]

– (1962): *Die alten Griechen und wir*, Göttingen.

Soden, Wolfram von (1965): *Leistung und Grenzen Sumerisch-Babylonischer Wissenschaft*, Darmstadt.

Spinner, Helmut (1977): *Begründung, Kritik und Rationalität. I: Die Entstehung des Erkenntnisproblems im griechischen Denken und seine klassische Rechtfertigungslösung aus dem Geiste des Rechts*, Braunschweig.

Sprat, Thomas (1667): *The History of the Royal Society*, ed. Jackson J. Cope and Harold W. Jones, St. Louis, 1958.

Sprenger, Jakob and Heinrich Institoris (1486): *Malleus Maleficarum*, trans. Montague Summers, London, 1928. [*Translator's note: "Institoris" is the Latinized version of "Kramer," which was the real name of the second author. The book was written in 1486 and first published in 1487.*]

Staudacher, Willibald (1942): *Die Trennung von Himmel und Erde ein vorgriechischer Schöpfungsmythos bei Hesiod und den Orphikern*, Tübingen.

Stesichorus [*Fragments*]: "Stesichorus: Fragments," in: D. A. Campbell (trans. and ed.), *Greek Lyric. Vol. III: Stesichorus, Ibycus, Simonides, and Others*, Cambridge, MA, 1991.
Stewart, John Alexander (1905): *The Myths of Plato*, New York, 1970.
Szabó, Árpád (1969a): *Anfänge der griechischen Mathematik*, Munich.
– (1969b): *The Beginnings of Greek Mathematics*, Dordrecht, 1978.
Thom, Alexander (1966): "Megalithic Astronomy: Indications in Standing Stones," *Vistas in Astronomy*, 7, pp. 1–56.
– (1967): *Megalithic Sites in Britain*, Oxford.
– (1969): "The Lunar Observatories of Megalithic Man," *Vistas in Astronomy*, 11, pp. 1–29.
– (1971): *Megalithic Lunar Observatories*, Oxford.
Thom, Alexander, Archibald S. Thom, and Alexander S. Thom (1975): "Stonehenge as a Possible Lunar Observatory," *Journal for the History of Astronomy*, 6, 15, pp. 19–30.
Thomson, Donald F. (1939): "The Seasonal Factor in Human Culture: Illustrated from the Life of a Contemporary Nomadic Group," *Proceedings of the Prehistoric Society*, 5, pp. 209–21.
Tylor, Edward Burnett (1873): *Religion in Primitive Culture*, New York, 1958.
Ucko, Peter J. and Andrée Rosenfeld (1967a): *Felsbildkunst im Paläolithikum*, Munich.
– (1967b): *Paleolithic Cave Art*, New York.
Ventris, Michael and John Chadwick (1956): *Documents in Mycenean Greek: Three Hundred Selected Tablets from Knossos, Pylos and Mycenae with Commentary and Vocabulary*, Cambridge.
Voltaire (1733): "Letter XIV: On Descartes and Sir Isaac Newton," in: *Letters Concerning the English Nation*, ed. and trans. Nicholas Cronk, Oxford, 1999.
Waerden, Bartel Leendert van der (1966): *Science Awakening. Vol. II: The Birth of Astronomy*, trans. Peter Huber, Oxford, 1974.
Webster, Thomas B. L. (1957): "From Primitive to Modern Thought in Ancient Greece," in: Fédération Internationale des Associations d'Études Classiques (ed.), *Acta Congressus Madvigiani. Vol. II: The Classical Pattern of Modern Western Civilization: Formation of the Mind, Norms of Thought, Moral Ideas*, Copenhagen, pp. 29–43.
– (1958): *From Mycenae to Homer*, London.
Weld, Charles R. (1848): *A History of the Royal Society: With Memoirs of the Presidents*, London.
Whitman, Cedric Hubbell (1958): *Homer and the Heroic Tradition*, Cambridge.
Whorf, Benjamin Lee (1956): *Language, Thought, and Reality: Selected Writings*, Cambridge.
Wilamowitz-Moellendorf, Ulrich von (1896): *Aischylos, Orestie: Griechisch und deutsch. Zweites Stück: Das Opfer am Grabe*, Berlin.
– (1914): *Aischylos Interpretationen*, Berlin.
– (1931): *Der Glaube der Hellenen*, Darmstadt.
Wissowa, Georg (1958): *Paulys Realencyclopädie der classischen Altertumswissenschaft*, ed. Konrad Ziegler with contributions from numerous colleagues in the field, Munich.
Wright, John K. (1925): *The Geographical Lore of the Time of the Crusades: A Study in the History of Medieval Science and Tradition in Western Europe*, New York.

Wulff, Oskar (1927): *Die Kunst des Kindes: Der Entwicklungsgang seiner zeichnerischen und bildnerischen Gestaltung*, Stuttgart.

Wundt, Wilhelm (1920): *Völkerpsychologie: Eine Untersuchung der Entwicklungsgesetze von Sprache, Mythos und Sitte*, vol. 4: *Mythos und Religion*, Leipzig.

Xenophanes [*Fragments*]: *Fragments: A Text and Translation with a Commentary*, trans. James Lesher, Toronto, 2001.

INDEX

general relativity theory 84
geometry
  and Descartes' approach to nature
    170
  Euclidean 83, 154
  and Hesiod's worldview 125
  and intuition 154–5
Geryon myth 35
Gigon, Olof 100, 140
Gilbert Islands 32
Gilbertese sky dome 13
God
  and Bacon's reform of knowledge
    177
  and the mechanistic worldview 195
  Xenophanes' concept of 143–7,
    156–7
  see also deities (ancient Greek)
Godelier, Maurice 77
Goodall, Jane 7
gravity
  and form 160
  Newton's theory of 192–3
  and special relativity theory 197,
    198, 199
Graziosi, Paolo 53
Greek antiquity xvii–xviii, xxii
  archaic art 51–64
  Bacon's assessment of 177
  emergence of philosophy in 87–112
  myth in xii, xxiii, 33, 34–45
  navigation 32
  Nietzsche on Greek mythology
    77–8
  and oral traditions 23
  theories of myth in 34–45
  see also Homeric epics; Pre-Socratic
    philosophy; preliterate cultures
Greek language
  and the development of Western
    philosophy of nature 92–3, 94–6
  and Homeric poets 66–8, 94–5
Greek tragedy
  and the emergence of philosophy
    101–3
  and Homeric epics 96
Groenewegen-Frankfort, Henriette
  A. 59
Guatemala Indians 11–12
Guthrie, William K. C.

A History of Greek Philosophy
  118, 147

Hades (the underworld) 120–4, 125,
  133
Hamlet's Mill (de Santillana and von
  Dechend) 42–3
Hanfmann, G. M. S. 61
Hart, Heinrich 85
Hawkins, Gerald S. 15
Hecataeus of Miletus 35, 88, 100–1
  world map 126, 127
Hegel, Georg Wilhelm Freidrich xxiv,
  40, 202
  and Bohr 202
  and the dynamics of concepts 169,
    185–92
  and the mechanistic worldview
    196, 197
  Science of Logic 187
Heisenberg, Werner xv, 162, 215
Helmholtz, Hermann von 195
Heraclitus 36, 88, 105, 206
Herder, Johann Gottfried 12
Herodotus 126
  Histories 101, 143
Hesiod 28, 34, 50, 87, 91, 137
  and Homeric epics 96
  Theogony 104, 105, 107–12, 125
  worldview 114, 115, 116, 117,
    120–5, 133, 134, 136, 141
Hindu astronomers 13
Hipour (Carolinian navigator) 14–15
history
  and the rise of modern science
    168–9
Hoddis, Jakob van
  End of the World 69
Höffding, Harald 199
Holz, Arno 85
Homeric epics xix–xx, xxi, 50, 64
  and Anaximander's worldview 118
  and cosmology 89, 141
  elimination of the Homeric
    worldview 89–90
  and the emergence of philosophy
    87, 94–5, 96–8
  event sequence and structure 79–81
  formulas in 65
  and Hesiod's worldview 110